bnas

W9-CSP-243

Cryptosporidium: The Analytical Challenge

Cryptosporidium: The Analytical Challenge

Edited by

M. Smith
Drinking Water Inspectorate, London, UK

K.C. Thompson
Alcontrol Laboratories, Rotherham, UK

ROYAL SOCIETY OF CHEMISTRY

Chemistry Library

UNIV. OF CALIFORNIA
WITHDRAWN

The proceedings of the two meetings on *Cryptosporidium: The Analytical Challenge* held on 25–26 October 1999 at the University of Warwick. The meeting was organised by the Water Chemistry Forum of the Royal Society of Chemistry together with the Environment and Water Group of the Society of Chemical Industry, the Chartered Institute of Water and Environmental Management (CIWEM) and the Drinking Water Inspectorate.

Special Publication No. 265

ISBN 0-85404-840-5

A catalogue record for this book is available from the British Library

© The Royal Society of Chemistry 2001

All rights reserved.

Apart from any fair dealing for the purpose of research or private study, or criticism or review as permitted under the terms of the UK Copyright, Designs and Patents Act, 1988, this publication may not be reproduced, stored or transmitted, in any form or by any means, without the prior permission in writing of The Royal Society of Chemistry, or in the case of reprographic reproduction only in accordance with the terms of the licences issued by the Copyright Licensing Agency in the UK, or in accordance with the terms of the licences issued by the appropriate Reproduction Rights Organization outside the UK. Enquiries concerning reproduction outside the terms stated here should be sent to The Royal Society of Chemistry at the address printed on this page.

Published by The Royal Society of Chemistry,
Thomas Graham House, Science Park, Milton Road,
Cambridge CB4 0WF, UK
Registered Charity No. 207890

For further information see our web site at www.rsc.org

Printed by MPG Books Ltd, Bodmin, Cornwall, UK

Preface

QL
368
C59
C79
2001
CHEM

This book constitutes the proceedings of the international conference held at Warwick University from 24th to 26th October 1999 entitled *Cryptosporidium*: The Analytical Challenge. It attempts to highlight the problems associated with this complex analysis and also indicates various means of overcoming them. The key (and often overlooked) issue of 'what is an oocyst' is comprehensively addressed.

It is a sobering thought that the new UK *Cryptosporidium* regulations require a minimum 30% recovery efficiency and a typical sample size of 1 tonne (1m³) of water sampled over a 24 hour period. Thus, it is required to be able to detect three oocysts in this quantity of tap water. This equates to a final mass concentration of 0.0002 ng litre^{-1} (parts per trillion), for a microbiological species that cannot be induced to replicate under the test conditions. This highlights one of the reasons for some of the problems experienced in this analysis.

The evolution of the UK *Cryptosporidium* regulations are outlined and the development of the official protocol for both the sampling and analysis steps are covered in detail. Comparisons are made with existing methodology.

Some novel techniques for the isolation, clean up and detection of *Cryptosporidium* oocysts are presented. Also some interesting work on the determination of species and genotype is given. The controversial issue of infectivity is discussed and both *in vivo* and *in vitro* methods are described.

After the isolation, concentration and clean up stages, the skills of the microscopist in correctly identifying *Cryptosporidium* oocysts from all other similar bodies is crucial. Mis-identifications have been made in the past with very significant consequences. False negatives or positives can have devastating effects for water companies. Much useful information is given in this key area.

The results for a *Cryptosporidium* Proficiency Scheme that has been running for six years are reviewed in depth and these clearly highlight the problems associated with this complex analysis. It was found that a group of laboratories in regular communication could improve their analytical performance significantly.

This book clearly demonstrates that although there is a considerable accumulated knowledge bank on the determination of *Cryptosporidium* in water, there are also many areas relating to this ubiquitous protozoan parasite that remain to be elucidated.

A subsequent conference scheduled for March 2000 will discuss the first year's experience in implementing the new UK *Cryptosporidium* regulations and hopefully should clearly demonstrate that the new regulatory sampling and analysis protocol is the correct way forward.

K. Clive Thompson, ALcontrol Laboratories
Mark Smith, Drinking Water Inspectorate
2nd July 2001

Contents

CRYPTOSPORIDIUM: THE ANALYTICAL CHALLENGE

*H.V. Smith and A. Ronald

Scottish Parasite Diagnostic Laboratory, Stobhill Hospital, Glasgow G21 3UW, UK.
*corresponding author. Tel: +44 (0)141 201 3028, Fax: +44 (0)141 558 5508

1 INTRODUCTION

In the last twenty years, the protozoan parasite, *Cryptosporidium* has been widely recognised as a cause of waterborne disease. Its transmissive stage, the oocyst, is a frequent inhabitant of raw water sources used for the abstraction of potable water and its importance is heightened because, coupled to its low infectious dose, conventional water treatment processes, including chemical disinfection, cannot guarantee to remove or destroy oocysts completely. Furthermore, because of their chlorine insensitivity, the coliform standard cannot be relied upon as an indicator of either the presence or viability of waterborne *Cryptosporidium* oocysts. For these reasons, robust, sensitive and specific methods are required for the recovery and identification of oocysts in water concentrates.

2 THE PARASITE

Cryptosporidium has a complex life cycle, involving both asexual and sexual reproductive cycles, which it completes within an individual host (monoxenous). Transmission from host to host is via an environmentally robust oocyst which is excreted in the faeces of the infected host. *C. parvum* is the major species responsible for clinical disease in man and domestic mammals (Current and Garcia, 1991), although infections with species other than *C. parvum* have been described in both immunocompetent and immunocompromised human hosts (Morgan *et al.*, 2000; Pedraza-Diaz *et al.*, 2000, 2001).

By the nature of its characters, the genus *Cryptosporidium* (Kingdom, Protozoa; Phylum, Apicomplexa (Sporozoa); Class, Coccidea; Order, Eimeriida) belongs to the Family Cryptosporidiidae. These include: "development just under the surface membrane of the host cell or within its striated border and not in the cell proper. Oocysts and meronts with a knob-like attachment organelle at some point on their surface. Oocysts without sporocysts, with four naked sporozoites. Monoxenous. Microgametes without flagella" (Levine, 1973). Currently, the genus *Cryptosporidium* contains 10 valid named species. *C. parvum*, *C. muris*, *C. felis*, *C. andersoni* and *C. wrairi* infect mammals, *C. baileyi* and *C. meleagridis* infect birds, *C. serpentis* and *C. saurophilum* infect reptiles and *C. nasorum* infect fish.

C. parvum infects the epithelia of the intestinal tract (enterocytes) of various mammals. Exposure to the environments of the gastrointestinal tract trigger the poorly understood process of excystation, whereby sporozoites are actively released through the suture of the oocyst. Known triggers include temperature (37°C), acidity (~ pH 2), slight alkalinity (~ pH 7.6) exposure to bile salts and trypsin (Fayer and Leek, 1984; Reducker and Speer, 1985; Hill *et al.*, 1991; Robertson *et al.*, 1993a). During the course of symptomatic infection, up to 10^{10} oocysts are shed into the environment, which are capable of prolonged survival in moist microclimates. Figure 1 describes the life cycle of *C. parvum*.

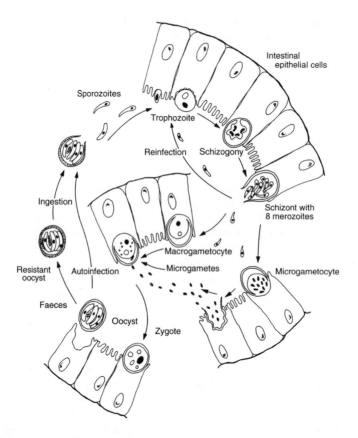

Figure 1 *Life cycle of* Cryptosporidium Parvum
Reprinted from *Parasitology Today*, **6**, H.V. Smith and J.B. Rose, 'Waterbourne Cryptosporidiosis', 8–12, 1990, with permission from Elsevier Science.

Waterborne transmission is well documented (Smith and Rose, 1990, 1998; Smith and Nichols, 2001), and can affect numerous individuals. Smith and Rose (1998) stated that more than an estimated 427,100 individuals had been affected in 19 documented waterborne outbreaks. The numerous contributors to waterborne oocysts, and the small size, environmental robustness and chlorine insensitivity of oocysts are factors that enhance their presence and likelihood of surviving in water treatment processes. The second report of the UK Joint Group of Experts on *Cryptosporidium* in water supplies stated that " ...

absence of *Cryptosporidium* from drinking water can never be guaranteed ..., criteria for 'best practice' for operating water treatment works could be identified and should be adopted" (Anon., 1995a). Well operated treatment works can achieve better than 2 log.$_{10}$ of oocyst removal and for disinfectants which could be used within current UK guidelines or legislation, ozone and UV have potential (Hall and Pressdee, 1995; Anon., 1995; Croll and Hall, 1997).

3 GETTING TOUGH ON *CRYPTOSPORIDIUM* OOCYSTS – A MULTINATIONAL APPROACH

Many requirements, rules and regulations are in place to attempt to address the control of *Cryptosporidium* which threatens the safety of drinking water. The current European Union 'drinking water' directive (Anon., 1980) requires that 'water intended for human consumption should not contain pathogenic organisms' and 'nor should such water contain: parasites, algas, other organisms such as animalcules'. The proposed revision to this directive, to be implemented in 2003 (Anon., 1995b), in recognising the impracticality of the current zero standard, will make it a general requirement 'that water intended for human consumption does not contain pathogenic micro-organisms and parasites in numbers which constitute a potential danger to health'. No numerical standard for *Cryptosporidium* is proposed.

The Unites States Environmental Protection Agency (USEPA) has issued several rules. One goal of the Surface Water Treatment Rule (USEPA, 1989) was to minimize waterborne disease transmission to levels below an annual risk of 10^{-4} by reducing *Giardia* cysts and viruses by 99.9% and 99.99%, respectively through filtration and disinfection requirements. The Rule lowered the acceptable limit for turbidity in finished drinking water to a level not to exceed 0.5 nephelometric turbidity unit (NTU, established by a standard haze created chemically in water and measured by light scatter) in 95% of four-hour measurements. The Enhanced Surface Water Treatment Rule (USEPA, 1994) includes regulation of *Cryptosporidium*, and in order to implement it, the Information Collection Rule (a national database on the occurrence of oocysts in surface and treated waters) had to be enacted.

In England and Wales, the Water Supply (Water Quality) Regulations 2000, SI No. 3184 and Water Supply (Water Quality) (Amendment) Regulations 1999, SI No. 1524, require water undertakers to determine whether there is a significant risk from *Cryptosporidium* oocysts in water supplied from waterworks and to comply with the requirement for treating the water intended to be supplied so that the average number of *Cryptosporidium* oocysts per 10 litres of water is less than one. Within the wording of the regulation, contravention of the numerical standard or the monitoring (sampling and analysis) requirements is an offence. To implement the regulations, which came into force on 30th June 1999, an information letter from the UK Drinking Water Inspectorate (DWI) (DWI Information letter 10/99) identified the Protocol containing Standard Operating Protocols (SOPs) for monitoring *Cryptosporidium* oocysts in water supplies. Similar directives are in place in Scotland and Northern Ireland. From data reported up to March 2001 under this regulation, 3,840 (8.8 %) of 43,740 samples taken from 178 water treatment works contained oocysts. Oocysts were detected in 43% of water treatment works which came under the regulation. Of the 3,840 positive samples, 500 (1.2 %) contained > 1 oocyst 100l^{-1} and 10 (0.023 %) contained > 10 oocysts 100l^{-1} (Drury, 2001).

In addition, the *Cryptosporidium* problems in Australia (McClellan, 1998) also highlight the need for robust methods, fit for propose, including detection, identification, enumeration and assessment of viability / infectivity of waterborne *Cryptosporidium* oocysts.

4 WHAT IS AN OOCYST?

The laboratory diagnosis of *Cryptosporidium* in faeces and its detection in environmental samples (e.g. water and food) is dependent upon demonstrating oocysts in the sample by microscopy. Here, oocysts must be distinguished from other, similarly shaped, contaminating bodies present in the sample, and the microscopic identification of oocysts is dependent upon morphometry (the accurate measurement of size) and morphology. The morphometric and morphological features of *C. parvum* oocysts viewed in suspension by Nomarski differential interference contrast (DIC) microscopy are as follows: spherical or slightly ovoid, smooth, thick walled, colourless and refractile, measuring 4.5 - 5.5 µm, containing four elongated, naked (i.e. not within a sporocyst(s)) sporozoites and a cytoplasmic residual body within the oocyst. Each sporozoite contains one nucleus. Not all four sporozoites may be visible at once. The size ranges of oocysts of various *Cryptosporidium* species are presented in Table 1.

Table 1 *Differences between some species within the genus* Cryptosporidium

Species of *Cryptosporidium*	Dimensions of oocysts (µm)	Site of infection	Host
Cryptosporidium parvum	4.5 x 5.5	small intestine	mammals
Cryptosporidium muris	5.6 x 7.4	stomach	mammals
Cryptosporidium andersoni	5.0-6.5 x 6.0-8.1	stomach	cattle
Cryptosporidium felis	4.5 x 5.0	small intestine	felids
Cryptosporidium wrairi	4.0-5.0 x 4.8-5.6	small intestine	guinea pigs
Cryptosporidium baileyi	4.6 x 6.2	trachea, bursa of Fabricius, cloaca	gallinaceous birds
Cryptosporidium meleagridis	4.5-4.0 x 4.6-5.2	intestine	turkeys
Cryptosporidium serpentis	4.8-5.6 x 5.6-6.6	stomach	snakes
Cryptosporidium saurophilum	4.2-5.2 x 4.4-5.6	intestinal & cloacal mucosa	lizards
Cryptosporidium nasorum	3.6 x 3.6	intestine	fish
Cryptosporidium sp.	4.5-6.0 x 3.6-5.6	small intestine	bobwhite quail
Cryptosporidium sp.	5.8-5.0 x 8.0-5.6	? intestine	snakes, reptiles

Detection in both clinical and environmental laboratories is performed on samples that have been dried down onto microscope slides or membranes. The drying process causes oocysts to collapse, which leads to shape distortion and the mechanical release of sporozoites (Figure 2). Commercially available, fluorescein isothiocyanate (FITC)-labelled monoclonal antibodies (mAbs; FITC-*C*-mAb), assist in the detection and measurement of oocysts in environmental samples by binding to surface exposed oocyst epitopes, hence defining shape (Rose *et al.*, 1989; Smith *et al.*, 1989a, Smith, 1995, 1996, 1998). The use of the fluorogenic dye 4'6 diamidino-2-phenyl indole (DAPI), which highlights sporozoite nuclei by fluorescing sky blue when intercalated with sporozoite DNA, provides supportive evidence (Grimason *et al.*, 1994; Smith, 1995, 1996, 1998).

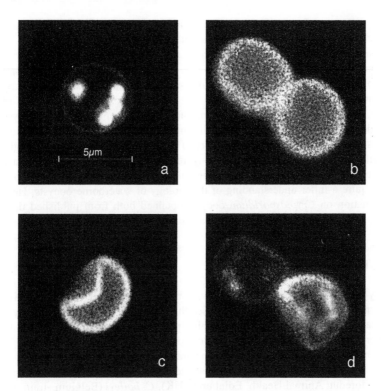

Figures 2a - d *Shape differences in C. parvum oocysts labelled with commercially available FITC-C-mAb. Magnification x1250*
Figure 2a *Intact C. parvum oocyst labelled in suspension with commercially available FITC-C-mAb and DAPI. Note the FITC fluorescence on the outer perimeter of the oocyst and the DAPI fluorescence of the 4 sporozoite nuclei*
Figure 2b *Two intact C. parvum oocysts labelled in suspension with commercially available FITC-C-mAb. Note the FITC fluorescence on the outer perimeter of the oocyst and the difference in their shapes*
Figure 2c *Intact C. parvum oocyst labelled in suspension with commercially available FITC-C-mAb. Note the FITC fluorescence on the outer perimeter of the oocyst and its concavo-convex shape*
Figure 2d *Two air dried, methanol fixed C. parvum oocysts labelled with commercially available FITC-C-mAb and DAPI. Note the FITC fluorescence on the outer perimeter of the oocyst, the DAPI fluorescence of the sporozoite nuclei and the variation in shape*

Occurrence studies of waterborne *Cryptosporidium* oocysts in numerous countries (Smith and Rose, 1998) have been performed using a limited (c. 3-4) number of commercially available (diagnostic) FITC-*C*-mAbs, which is indicative of the fact that the epitopes recognised by these FITC-*C*-mAbs are not only commonly expressed in *Cryptosporidium* species oocysts, but are also environmentally robust. For many in the water industry, interest in *Cryptosporidium* oocysts is rarely required to extend beyond their identification in water concentrates, where reliance is placed upon the reactivity of FITC-*C*-mAb, DAPI, knowledge of oocyst morphology and experience, yet a clearer understanding of oocyst form and function may lead to better detection and destruction methods which, in turn could reduce the risk associated with waterborne oocysts.

C. parvum oocysts afford a stable and resistant covering for infective sporozoites, providing protection from external influences (adverse temperatures, desiccation, salinity, disinfection processes and other environmental insults) yet remaining sensitive to triggers known to initiate excystation and the active release of sporozoites. Knowledge of the ultrastructure, physiology, biochemistry, biophysics and antigenicity of the oocyst can assist our understanding of this environmentally robust, protective structure. In addition, a clearer understanding of the nature and structure of the oocyst can also provide insight into improved methods for identification and sub-typing by targeting novel markers, and for disinfection, by identifying structures and topographies sensitive to disinfectants. For these reasons we require a fuller understanding of the biology of waterborne oocysts. Here, we present information on *Cryptosporidium* oocysts obtained both from published data and experiments performed at the SPDL over the last ten years, with particular reference to the oocyst wall.

5 OOCYST SIZE AND MORPHOLOGY

Morphometry and morphology frequently provide the basis for the decision as to whether an object is an oocyst. From Table 1 we can see that there is overlap between oocyst size ranges for many of the species presented, which presents a significant limitation to our ability to determine the presence of *C. parvum* oocysts at the light microscope level.

To determine the degree of overlap between similarly sized oocysts of *Cryptosporidium* species likely to contaminate UK waters, we obtained stocks of *C. parvum* (cervine / ovine isolate, MD, Moredun Animal Health, Edinburgh, UK), *C. baileyi* (Belgium strain, LB 19; Dr. K. Webster, Veterinary Laboratories Agency, Weybridge, UK) and *C. muris* (RN 66, Dr. V. McDonald, London School of Hygiene and Tropical Medicine, London, UK) oocysts and stored them in reverse osmosis (RO) water containing antibiotics at 4°C. We measured oocysts suspended in RO water, using an Olympus BH2 microscope equipped with DIC optics at a total magnification of x1250 (12.5 eyepieces x 100 oil objective). Forty oocysts each of *C. parvum*, *C. muris* and *C. baileyi* were measured (Table 2).

Table 2 *Differences in oocyst size between* C. parvum, C. baileyi *and* C. muris

Cryptosporidium species	Oocyst dimensions (µm ± s.d. of population; n = 40)
C. parvum	5.0 0 ± 0.35 x 5.45 ± 0.47
C. baileyi	5.08 ± 0.19 x 6.30 ± 0.52
C. muris	6.40 ± 0.39 x 8.54 ± 0.66

Given a maximum size range of 5.35 x 5.9 for *C. parvum* oocysts in this study, analysis of the minimum size of *C. muris* oocysts (6.25 x 7.5 µm) indicate that they are larger than the maximum size range for *C. parvum* oocysts. Therefore, based on morphometry, intact *C. muris* oocysts can be distinguished readily from *C. parvum* oocysts in water concentrates. The minimum size of *C. baileyi* oocysts detected was 5.0 x 5.0 µm which falls within the size range for *C. parvum* oocysts and analysis of our data indicate that 40% (16/40) of the *C. baileyi* oocysts measured fall within the maximum size of *C. parvum* oocysts.

If we adopt the morphometric guidelines for *C. parvum* oocysts identified in the provisional recommended UK methods and the UK Regulatory method (Anon., 1990, 1999a,b; 4-6 µm in size) then 45% (18/40) of our *C. baileyi* oocysts fall within this size range. Using the dimensions of Tyzzer (1912) and Upton and Current (1985) (4.5 x 5.0 µm), then 10% (4/40) of our *C. baileyi* oocysts fall within this size range. Immaterial of which size range we adopt, there is overlap between the measured sizes of *C. parvum* and *C. baileyi* oocysts. These data indicate that morphometric measurements will not always be sufficient to discriminate between oocysts of *C. parvum* and *C. baileyi*, therefore better discriminatory methods are required. The occurrence of other *Cryptosporidium* spp. oocysts which are similar in size to *C. parvum* oocysts (Table 1) in the aquatic environment, including human-derived *C. meleagridis* oocysts, further compromises size-based detection methods.

6 *C. PARVUM* GENOTYPES

Recent genetic analyses have raised doubt about the validity of current species, and previously accepted criteria including oocyst morphology, host specificity and parasite location may not be sufficiently discriminatory (Tzipori and Griffiths, 1998). Analysis of several polymorphic sites in at least 6 different genetic loci of the *C. parvum* genome (e.g. *Cryptosporidium* oocyst wall protein (COWP), dihydrofolate reductase, *Cryptosporidium* thrombospondin related adhesive protein-1 & -2, ribonuclease reductase and the internal transcribed spacer 1 of the 18s rRNA gene) indicate that *C. parvum* is composed of two distinct genotypes: genotype 1, which infects humans, primarily, and genotype 2,which infects both humans and other mammals, particularly ruminants and rodents.

7 OOCYST WALL STRUCTURE

The environmental robustness of *Cryptosporidium* oocysts is well documented. At the host-oocyst-environment interface is the oocyst wall, which is often regarded as a near-inert protective covering.

The *C. parvum* oocyst wall is between 40 nm (Harris and Petry, 1999) and 50 nm (Nanduri *et al.*, 1999) thick, although Reduker *et al.* (1985a) cite a range from 31.6 to 72.9 (mean, 49.7 ± 11.5 nm). At the outer oocyst wall is a carbohydrate-rich glycocalyx, 20 to 30 nm thick, containing traces of a C18 fatty acid (Nanduri *et al.*, 1999). The oocyst wall possesses both outer and inner layers. The inner layer can be further subdivided into outer-inner and inner-inner layers (see Figure 3). Reduker *et al.* (1985a) found that the outer-inner layer was of irregular thickness (mean, 10 nm) and the inner-inner layer comprised an outer zone (mean, 11.6 nm) and an inner zone (mean, 25.8 nm).

The outer layer is between 5 and 10 nm thick and variably electron dense, with sparse filamentous material extending outwards from the oocyst surface. On negatively stained dried oocyst sections, the outer rim of the flattened wall appears as an electron transparent edge, characteristic of a double layer structure, as seen in negatively stained membrane vesicles, erythrocyte ghosts and nuclear envelopes (Harris and Petry, 1999). The outer layer of the wall is thought to be composed of acidic glycoprotein with an unknown amount of lipid (glycolipid / phospholipid) either loosely or firmly associated with the outer wall (Harris and Petry, 1999; Robertson *et al.*, 1993b).

The ~ 5nm thick central layer is composed of complex lipid, possibly responsible for the acid-fast staining of the oocyst wall, and is thought to provide a large proportion of the oocyst's apparent rigidity (Harris and Petry, 1999).

The inner layer is composed of a 10 nm electron dense layer and 20 nm layer which is less electron dense and contains a suture which is up to a half of the oocyst's circumference. The 10 nm electron dense layer shows signs of a particulate structure typical of a fibrillar array, which is clearly visible as a disordered filamentous material in negatively stained sections. The inner, 20 nm, layer is composed of a filamentous glycoprotein (Bonin *et al.*, 1991), susceptible to proteinase K and trypsin, but not pepsin digestion, and may provide much of the rigidity and elasticity present in intact oocyst walls (Harris and Petry, 1999). Oocyst wall components are thought to be transported to the surface of the developing oocyst in Type 1 and Type 2 wall forming bodies (Bonin *et al.*, 1991; McDonald *et al.*, 1995; Spano *et al.*, 1997).

8 *CRYPTOSPORIDIUM* OOCYSTS: MOLECULES AND ANTIGENS

Molecular and antigenic studies are frequently complementary, particularly for proteins, carbohydrates and glycoconjugates. While many of the studies presented have used aqueous extracts of whole oocysts, some information is also available on molecules and antigens localised to the oocyst wall.

8.1 Lipids

Many lipid species are present in *C. parvum* oocysts. White *et al.* (1997) found different types of both polar and neutral lipids and a number of different glycolipid species. Not only were differences in the ratios of these different lipid species noted between *C. parvum* and *C. muris* oocysts, but also a noticeable variation in the ratios of these lipids occurred in viable and non-viable oocysts, with a decrease in the levels of cholesterol in *C. parvum* oocysts killed by freezing. That the nature and quantity of lipid species in oocysts can be used to distinguish between live and dead oocysts is an indication of their importance to the integrity and survival of the oocyst. Schrum *et al.* (1997) raised the possibility of utilising 'signature' lipid bio-markers, not only to distinguish between oocyst species but also to determine live from dead oocysts. Unfortunately, the 'signature' lipid biomarker chosen (10-hydroxy stearic acid) turned out to be an artefact of the preparation procedure (Burkhalter *et al.*, 1998).

However, evidence exists for a correlation between loss of infectivity and oocyst fatty acid and lipid content. There is a direct correlation between *C. parvum* oocyst amylopectin content and infectivity, together with more than a threefold decrease in polar lipid fatty acids (PLFA) from 3.8 to 0.75 pmoles 10^3 oocyst^{-1} and a change in the patterns of lipids present, including a decrease in the proportion of 18:1n-9c and an increase in 18:2n-6 and 20:3n-6 (Burkhalter *et al.*, 1998). A consequence of freezing oocysts is the decrease in phospholipids and cholesterol and the increase in polyenoic PLFA, which may be due to selective loss of lipid residues during cooling. These lipid markers may prove useful in differentiating between infective and non-infective oocysts, but can only be used with relatively pure oocyst preparations as a number of different algae and bacteria overlap in lipid composition. PLFA patterns of purified oocysts can be used to distinguish between *C. parvum* and *C. baileyi*, *C. serpentis* and *C. muris* (Burkhalter *et al.*, 1998).

Treatment of oocysts with ß-cyclodextrin, which induces erythrocyte haemolysis by removing membrane components such as cholesterol and phospholipids, leading to membrane dissolution (Ohtani *et al.*, 1989), led to a marked reduction in oocyst viability (Castro-Hermida *et al.*, 2000). *C. parvum* oocysts contain various lipids, including phospholipid and sterols, and Mitschler *et al.* (1994) found that of approximately 1.2×10^{-9} μmol phospholipid present per oocyst, about 66% consisted of phosphatidylcholine. The only sterol detected was cholesterol, at approximately 1.7×10^{-10} μmol per oocyst.

Some oocysts exposed to ß-cyclodextrin and incubated with malachite green or DAPI and propidium iodide (PI), internalised these stains, indicating that alterations in the permeability of the oocyst wall had occurred, and that treatment with ß-cyclodextrin led to lipid loss (phospholipids and cholesterol) from the outer oocyst wall, which, in turn, increased oocyst permeability and reduced oocyst viability. Increased permeability leading to oocyst death occurred in only a proportion of the population (Castro-Hermida *et al.*, 2000), indicative of some heterogeneity in the population studied, possibly due to a mixture of clones and / or the oocyst purification procedure.

Studies based on analysis of extracted lipid provide limited information about their location in the oocyst. We argued for the presence of lipid in the *C. parvum* oocyst wall by demonstrating that the surface-reactive, lipophilic, cationic fluorescent dye, octadecyl rhodamine B (R18) bound surface-associated molecules of intact oocysts, highlighting the same topography as a commercially available FITC-*C*-mAb (Robertson *et al.*, 1993b). Further evidence for the presence of surface-associated lipid moieties came from our observation that the fluorogenic lipid analogue 5-(N-octadecanoyl) aminofluorescein (AF18) also inserted into the wall of intact oocysts. In the epicuticular surfaces of adult nematode parasites, selectivity of insertion of fluorescent lipid probes is thought to be primarily due to the physicochemical properties of the probe. The anionic AF18 probe is readily inserted into in the parasitic nematode surface, but the cationic probe octadecylrhodamine B (RH18), among others, is not (Proudfoot *et al.*, 1991). Since AF 18 and RH 18 have the same hydrophobicity, selectivity must be largely head group dependent, although, in the parasitic nematode surface, (Kennedy *et al.* (1987) found a strict dependence on acyl chain length within the aminofluorescein analogues.

Further evidence for the presence of lipid molecules in the outer layer of the oocyst comes from the work of Nanduri *et al.* (1999) who found trace amounts of a C-18 fatty acid in the predominantly carbohydrate glycocalyx on the outer surface of *C. parvum* oocysts, and Brush *et al.* (1998) who found that the use de-fatting agents and other clean up procedures to purify oocysts led to alterations in the surface charge of oocysts, presumably due to lipid (or other ether/ ethyl acetate soluble materials) removal from the oocyst wall.

Analysis of purified *C. parvum* oocyst wall preparations by thin layer chromatography revealed the presence of glycolipids, but not phospholipids while acid methanloysis and gas chromatography-mass spectroscopy analysis of extracted lipids revealed the presence of hexadecanoic, 6-hexadecanoic, octadecanoic, 9-octadecanoic (18:1n-9c) and 11,14 eicosadienoic acids (Anthony *et al.* 1998). Significant quantities of C16 to C21 aliphatic hydrocarbons were identified in hexane and chloroform-methanol extracts, while hexane extraction caused ultrastructural damage to the outer layers of oocyst walls. Anthony *et al.*, (1998) proposed that the presence of aliphatic hydrocarbons and lipids in the central, electron transparent layer of the oocyst wall might influence oocyst wall permeability.

Cryptosporidium oocysts contain lipid moieties, including large amounts of phosphatidylcholine, phosphatidylethanolamine, cholesterol, PLFA and a range of other polar and neutral fatty acids and phospholipids. Some of these are located in the outer layer of the oocyst wall, where they play an important role in determining the permeability of the oocyst wall, its surface charge and hydrophobicity, and, by extrapolation, the resistance of the oocyst to environmental and possibly disinfection processes.

8.2 Antigens

The antigenic nature of intact *C. parvum* oocysts has been investigated using both mAbs and polyclonal antibodies (pAbs) raised following immunisation of various oocyst antigen preparations or following experimental or natural infection of non-human and human hosts. McDonald *et al.* (1991) found that the majority of hybridomas raised against whole oocysts were reactive with sporozoite antigens and that <10% were reactive with the intact oocyst wall. Of the commercially available mAbs raised against exposed oocyst wall antigens, most are of the IgM isotype, suggestive of a carbohydrate epitope, which supports the concept of a glycosylated outer oocyst surface. Intact oocysts appear to be poorly immunogenic, and isolated oocyst walls elicit a poor immune response, reacting weakly in Western blots with a limited number of antigens (Weir *et al.*, 2000). Similarly, in our experiments, immunisation of mice with intact oocysts generated lower serum antibody titres compared with immunisation with purified excysted oocyst walls or sodium dodecyl sulphate (SDS) extracted oocyst surfaces. Serum antibody titres increased more slowly with intact oocysts. In addition, *C. parvum* oocysts, stripped with SDS, also generated poor murine immune responses following immunisation. In our experience, whole oocysts and SDS-stripped oocysts appear to be poor immunogens, possibly even immunosuppressive. Reducker *et al.* (1985) suggested that the low immunogenicity of isolated oocyst walls could be due to changes in the oocyst wall caused by excystation media, but Weir *et al.* (2000) disagreed because isolated oocyst walls remained reactive to diagnostic FITC-*C*-mAbs. Despite these limitations, mAbs and pAbs, reactive to both the outer and inner layers of the oocyst wall have been used to analyse the antigenic nature of *Cryptosporidium* spp. oocysts.

Some epitopes are distributed evenly on / in the *Cryptosporidium* oocyst wall, while others are unevenly distributed (Jenkins *et al.*, 1999). Diagnostic FITC-*C*-mAbs display a smooth, even fluorescence, indicative of an evenly distributed, high density epitope covering the outer oocyst surface, when reacted with intact oocysts in suspension or air dried onto microscope slides. FITC-*C*-mAb paratopes bind surface-exposed oocyst epitopes and the fluorescence visualised defines the maximum dimensions of the organism, enabling morphometry to be performed readily. In our experience, some mAbs, reactive with the intact *C. parvum* outer oocyst wall, display an uneven, granular or patchy fluorescence, indicative of more unevenly distributed epitopes with varying exposed densities, and, presumably for these reasons, never reach the diagnostic marketplace.

8.2.1 FITC-C-mAb epitope(s) and dual labelling. Although diagnostic FITC-*C*-mAbs are used routinely for *Cryptosporidium* analysis, worldwide, little is known of the oocyst wall epitope(s) to which they bind. We used two diagnostic mAbs (mAbs A and B) to analyse oocyst wall antigen profiles. Oocyst antigens were extracted from MD isolate oocysts using 2 different methods (Campbell *et al.*, 1993; Ronald *et al.*, 2001). Extracts were run on SDS-PAGE, Western blotted and probed with mAb A or mAb B. Probing MD oocyst wall antigens, extracted by boiling in SDS and ß-mercaptoethanol (mAb A) for 5 min revealed a total of 13 bands ranging from 17 – 220 kDa, while probing MD oocyst wall antigens, extracted by overnight incubation in SDS at 37°C (mAb B) revealed a total of 5 bands ranging from 57 – 117 kDa. Despite the different treatments, four bands (57, 66-

68, 75-77 & 110-117 kDa) were common to both mAbs (Table 3). Boiling oocysts in ß-mercaptoethanol is likely to release more oocyst antigens than overnight incubation in SDS at 37°C, resulting in a broader antigen profile. Incubation of antigens extracted by overnight incubation in SDS at 37°C in ß-mercaptoethanol failed to generate further antigen fragments. Boiling oocysts in SDS and ß-mercaptoethanol is more likely to extract antigens which are not oocyst wall-associated, whereas antigens extracted by overnight incubation in SDS at 37°C incubation are more likely to be surface-associated, as this treatment does not lead to oocyst rupture.

Table 3 *Comparison of two FITC-C-mAb reactivities with C. parvum (MD isolate) oocyst wall antigens*

mAb A (kDa)	mAb B (kDa)
220	
160	
132	
110	117
86.5	
75	77
68	66 - 68
	61 - 63
57	57
50	
41	
38	
22	
17	

Dual labelling experiments with MD isolate *C. parvum* oocysts were performed at the SPDL, using commercially available FITC-*C*-mAb C and biotinylated, commercially available mAb A (both IgM isotype; NHS-biotin, Hurley *et al.*, 1985). In a dual labelling (competition) assay, a suspension of MD isolate oocysts was incubated with FITC-C-mAb C and biotinylated mAb A, and incubated at 37°C for 30 min in the dark, then unbound mAbs were removed by washing. Streptavidin-TRITC, which bound to biotinylated mAb A, was added to the labelled oocysts and incubated at 37°C for 30 min in the dark, and unbound conjugate removed by washing. Neither mAb A nor mAb C labelled 100% of oocysts. 80% of oocysts dual labelled with both mAbs, but the remaining 20% stained with either one mAb or the other. Of these, 13% labelled with mAb C and 7% with mAb A (Campbell *et al.*, 1993a), which is indicative of population heterogeneity.

8.2.2 Sodium dodecyl sulphate-polyacrylamide gel electrophoresis (SDS-PAGE) and Western blot analyses.

<u>Inter and intra-species comparisons</u>

SDS-PAGE and Western blot analyses of *C. parvum* oocysts have revealed a complex pattern of antigens, ranging from <14 to >200 kDa (Luft *et al.*, 1987; Lumb *et al.*, 1988; McDonald *et al.*, 1991; Nina *et al.*, 1992a; Lorenzo *et al.*, 1993; Ortega-Mora *et al.*, 1992). The complexity and diversity of molecules present in *Cryptosporidium* oocysts is in direct contrast with the situation seen in oocysts of the related coccidian *Eimeria tenella*, which appears to consist of a single 10 - 12 kDa major oocyst wall protein antigen by Western blotting, although the presence of a faint 24kDa band may indicate that the natural form of this protein is dimeric or multimeric (Stotish *et al.*, 1978; Karim *et al.*, 1996).

Many mAbs raised against *C. parvum* oocysts cross react with other life cycle stages (Lumb *et al.*, 1989; Bjorneby *et al.*, 1990; McDonald *et al.*, 1995), other *Cryptosporidium* species (Chrisp *et al.* 1991; Nina *et al.* 1992a) and other coccidian parasites, indicating the existence of common epitopes. PAbs raised against *C. parvum* oocysts cross react with *E. tenella* oocysts by immunofluorescence, and sera from *Cryptosporidium*-naïve lambs, infected with *E. tenella*, reacted with multiple *C. parvum* oocyst antigens, ranging from 29 to 69 kDa by Western blotting (Ortega-Mora *et al.* 1992), indicating that the epitope of the 12 kDa *E. tenella* oocyst wall antigen is also present in *Cryptosporidium* oocyst walls. A range of mAbs raised against *C. parvum* oocysts cross react with both *C. parvum* and *C. wrairi* oocysts by immunofluorescence with similar intensities (Chrisp *et al.*, 1995) and pAbs recognise some antigens common to *C. parvum*, *C. baileyi* and *C. muris* oocysts (Nina *et al.* 1992b). Such studies indicate that major epitopes are highly conserved between *Cryptosporidium* spp. oocysts, despite differences in parasite biological behaviour and host specificities (Chrisp *et al.*, 1991).

Both the number and nature of oocyst wall antigens vary between and within species (McDonald *et al.*, 1991; Nina *et al.*, 1992a,b), with greater variation between different *Cryptosporidium* species than between different *C. parvum* isolates. Nina *et al.* (1992b) showed that while mAb 181B5 bound to oocysts of *C. parvum*, *C. baileyi* and *C. muris*, the intensity of binding was noticeably less to both *C. baileyi* and *C. muris* oocysts than to *C. parvum* oocysts. Furthermore, a panel of sporozoite-reactive mAbs also displayed differences in binding to the same panel of *C. parvum*, *C. baileyi* and *C. muris* oocysts. Western blotting, with anti-*C. parvum* or anti-*C. muris* antiserum, revealed differences in the antigenic profiles of these three species, with the anti-*C. parvum* sera detecting 30, 7 and 36 antigenic bands in *C. baileyi*, *C. muris* or *C. parvum* oocyst extracts, respectively, and 25, 18 and 22 bands with the *C. muris* antiserum. Using *C. muris* antiserum, a number of major bands were unique to each species: *C. baileyi* displayed antigens at 56 – 65 & 100 (doublet) kDa, *C. muris* at 51 - 54 & 110 kDa, and *C. parvum* at 48 & 58 kDa. With *C. parvum* antiserum, bands of 51 & 56 kDa and 38, 42 & 48 kDa were seen only in *C. baileyi* and *C. parvum* oocyst extracts, respectively.

Some antigenic differences between *C. parvum* isolates are relatively minor, such as subtle differences in the molecular mass of antigens recognised by mAbs (Nina *et al.* 1992a), yet others can be major. Lumb *et al.* (1988) described two distinct types of antibody reactivity to excysted, human-derived *C. parvum* oocyst wall antigens in human infection sera: the first included reactivity with common (23 & 32 kDa) and broad range (23 to > 200 kDa) antigens and the second, reactivity with fewer antigens (common 23 & 32 kDa as well as 40 -180 kDa).

Previous data obtained at the SPDL led us to believe that greater variability existed in the outer oocyst surface than had been previously reported. We found that SDS extracted oocyst surfaces from two commercially supplied isolates of the 'Iowa' strain, generated different antigenic profiles, with a major 45-46 kDa antigen being present in one, but absent from the other isolate (Ronald *et al.*, 2000). We investigated oocyst wall antigenic variability further by analysing the antigenic nature of oocyst surfaces of various *C. parvum* isolates, including genotypes 1 & 2 and isolates from different host species by SDS-PAGE and Western blotting, using both pAb and diagnostic FITC-*C*-mAb reagents. We stripped surface-associated molecules over an extended time period at 37°C, which was less damaging to both oocyst intactness and viability. Blots of SDS-oocyst wall extracts revealed a common antigenic profile for all genotype 1 *C. parvum* oocysts investigated (n = 9), but different and varying antigenic profiles for *C. parvum* genotype 2 oocysts, of both human and non-human origin (n = 20). A marked degree of variation was seen between different genotype 2 oocyst isolates, and was, at least partly, associated with the host species from which oocysts were isolated (Ronald *et al.*, 2001), which is suggestive of the acquisition and / or insertion of host antigens (e.g. glycoconjugates) into the oocyst wall.

Surface-associated molecules and antigens

Techniques that primarily label surface-reactive molecules can provide a clearer understanding of molecules that make up *Cryptosporidium* oocyst outer surfaces. Lumb *et al.* (1988) analysed the surface-reactive molecules of 5 *C. parvum* isolates, using a modified Bolton and Hunter reagent (N-hydroxysuccinimide ester) to ^{125}I radiolabel the *C. parvum* oocyst surface, SDS-PAGE and autoradiography. Of four human-derived and 1 cervine-derived isolates, common bands occurred at 15, 32, 42, 47.5, 79 & 96 kDa in all isolates tested. Bands at 52, 62, 116 & 132 kDa were also present in some, but not all, isolates tested. Western blots of purified oocyst walls revealed common antigens of 23 & 32 kDa, as well as antigens in the 40 -180 kDa range (Lumb *et al.*, 1988). Tilley *et al.* (1990) found that ^{125}I-labelled (Bolton and Hunter reagent) surfaces of *C. parvum, C. baileyi* and *C. muris* oocysts contained a broader range of surface-reactive molecules. SDS-PAGE profiles of ^{125}I-labelled *C. parvum* oocyst surfaces revealed 17 distinct bands, while both *C. baileyi* and *C. muris* oocyst profiles revealed 18 bands. Surface-associated molecules of 32, 57, 120, 145-148 & 285-290 kDa were common to all three species studied. *C. baileyi* and *C. muris* shared common bands at 18-19, 29, 80-81, 100 & 180 kDa. A 46-47 kDa band was common to *C. parvum* and *C. baileyi* and a 67-69 kDa band common to *C. parvum* and *C. muris*. A number of bands were also unique to each species. Tilley *at al.* (1990) reported their banding patterns to be similar to those of Lumb *et al.* (1988), with proteins of 16, 32 & 46-49 kDa corresponding to the Lumb *et al.* proteins of 15, 32 & 47.5 kDa, and the 73 kDa Tilley *at al.* band possibly matching the 79 kDa band reported by Lumb *et al.* However, the 42 and 96 kDa bands reported by Lumb *et al.* (1988) had no corresponding Tilley *at al.* homologue.

Lumb *et al.* (1988) observed differences between ^{125}I-labelled profiles (15.5, 32 & 47.5 kDa) and Western blotting profiles (23, 32, 40-180 kDa). The ^{125}I-labelled 47.5 kDa protein did not react with the human sera tested and the 23 kDa antigen did not label with Bolton and Hunter reagent. The 23 kDa antigen may not a) be surface-exposed or b) contain sufficient lysine residues for Bolton and Hunter iodination. Clearly, although there are similarities in these banding profiles, differences also exist.

*8.2.3 Immunolocalisation.*Studies localising antigens to *Cryptosporidium* oocyst walls are fewer in number. McDonald *et al.* (1995) demonstrated mAb 1B5 binding to intact *C. parvum* oocysts by immunofluorescence. Ultrastructurally, mAb 1B5 bound to macrogametocytes, particularly the electron dense bodies involved in oocyst wall formation (Current and Reese, 1986), microgametes, and both the inner and outer layers of the oocyst wall (McDonald *et al.* 1995), indicating that common antigens are shared between both the inner and outer oocyst wall. Using mAb 1B5, McDonald *et al.* (1991) and Nina *et al.* (1992a) identified two major antigen bands (41 & 44 kDa) in *C. parvum* oocysts, which were not present in either *C. muris* or *C. baileyi* oocysts. Antigens of 72 & 110 kDa were present in *C. muris* oocysts and antigens ranging from 35 to > 200 kDa were present in *C. baileyi* oocysts.

An oocyst wall reactive IgM antibody (OW-IGO), which binds to inner and outer layers of the *C. parvum* oocyst wall, recognises a 250 kDa oocyst antigen and several minor components by Western blotting, under non-reducing conditions (Bonnin *et al.*, 1991). In immuno-electron microscopy studies, OW-IGO labelled the vacuolar space around microgametes and, in early macrogametes, the epitope was present in and around electron-lucent vesicles, with increasing amounts detected as macrogametes developed. Immunolocalisation provided evidence for the involvement of macrogamete vesicles in the storage of a glycoproteinic antigen released in the vacuole and later incorporated into the oocyst wall. This glycoproteinic fibrillar material, present in both the inner and outer wall of thick-walled oocysts is also present in the more fragile, thin-walled oocysts, suggesting that the molecules characterised by OW-IGO may not be involved in the robustness of the oocyst (Bonnin *et al.*, 1991).

Ranucci *et al.* (1993) screened *C. parvum* genomic libraries with an oocyst antiserum and a specific gene probe and identified two overlapping clones containing an open reading frame encoding a 1,252 amino acid polypeptide. Analysis of the deduced amino acid sequence of this protein revealed unusually high amounts of cysteine, proline and histidine. Both immune sera and mAbs raised against a recombinant polypeptide encompassing the first 786 amino acids recognised a *C. parvum* protein with an apparent molecular mass of 190,000, and were used to investigate its expression and localization in *C. parvum*. Immunolocalisation was to the *Cryptosporidium* oocyst wall (hence *Cryptosporidium* oocyst wall protein (COWP)-190). The high cysteine content and amino acid repeats suggest that COWP-190 may play an important functional and / or structural role in the oocyst wall. Further analysis of the amino acid sequence revealed the presence of two different repeat motifs, Type I and Type II, which were divided into two major domains (Spano *et al.*, 1997). The amino-terminal domain, encompassing the first 698 amino acids consisted of a putative leader peptide followed by 10 tandemly-arrayed Type I repeats of approximately 65 amino acids each. The carboxyl-terminal domain encompassed 772 amino acids, containing 8 Type II repeats of 53 amino acids, alternating with histidine-rich sequences of variable length (Spano *et al.*, 1997). The high frequency of synonymous mutations that occur within these coding sequences indicates the presence of selective pressures to conserve the amino acid sequences of these repeats (Spano *et al.*, 1997).

MAbs raised against this recombinant protein reacted with a 190 kDa antigen from a *C. parvum* oocyst lysate in Western blots under reducing conditions, but under non-reducing conditions, no bands were seen, either with mAb or immune sera raised against the recombinant antigen. The high cysteine content was implicated in the lack of reactivity of the native protein, with cross-linking of these regions resulting in the native form existing as a large molecular aggregate, which was unable to enter the polyacrylamide gel (Spano *et al.*, 1997).

Immunogold ultrastructural localisation of this protein in ultra-thin sections of *C. parvum* oocysts clearly implicated COWP-190 as an abundant constituent of the inner wall of the oocyst. Macrogametes also contained the COWP-190 epitope. Localisation in early macrogametes was to a single electron dense vesicular body, and to smaller cytoplasmic granules (possibly Type I wall forming bodies) in more mature macrogametes. These data provide evidence for the development of Type I wall forming bodies from a larger vesicular precursor, and a direct antigenic relationship between Type I wall forming bodies and the inner wall of *Cryptosporidium* oocysts (Spano *et al* 1997). No immunolocalisation was seen in sporozoites, trophozoites or schizonts.

The abundant, cysteine rich COWP-190 protein may play an important role in the structural integrity and strength of the oocyst wall, with the amino acid repeats providing periodic spacing and appropriate conformation to the protein to allow the formation of a large number of disulphide bonds. This would enable COWP-190 to form an intra- and inter-molecular net of chemical bonds, conferring structural stability to the oocyst wall (Spano *et al.*, 1997).

8.2.4 Disulphide bonds. The importance of disulphide bonds in *Cryptosporidium* oocyst antigen structure has been demonstrated using reducing agents such as 2-mercaptoethanol and dithiothreitol, following SDS-PAGE and Western blotting. Under non-reducing conditions, Luft *et al.* (1987) found that wall antigens of excysted oocysts resolved into two high molecular mass bands, and as many as seven in total ranging from 55kDa to >100kDa, but 2-mercaptoethanol treatment consistently resolved them into antigens of 72, 76, 98 & >100 kDa. Similarly, McDonald *et al.* (1991) found that mAb 2B2 bound strongly to a high molecular mass component, too large to be resolved by 10% SDS-PAGE, a doublet of 190 kDa, and bound weakly to a 40 kDa antigen. Treatment with 2-mercaptoethanol markedly reduced the intensity of mAb 2B2 binding to the major unresolved antigen, but revealed a major 43 kDa band and a faint band at 40 kDa. MAb OW-IGO (Bonnin *et al.*, 1991) recognised a major 250 kDa antigen and several minor components under non-reducing conditions, but recognised a more complex, lower molecular mass antigen profile under reducing conditions, including the appearance of a major 40 kDa antigen, present only as a minor component under non-reducing conditions.

8.3 Carbohydrate moieties

8.3.1 Periodate oxidation and lectin reactivity. Many surface-associated molecules of parasites are glycosylated and serve various functions including protection, locomotion (slime), regulation of physiology, etc. Surface-associated carbohydrates are at the host-parasite interface and play an important role in the composition, structure and antigenicity of *C. parvum* oocyst walls. Studies on the antigenic composition of *C. parvum* have highlighted the role of carbohydrate moieties as significant epitopes (Ward and Cevallos, 1998). Antibody binding to carbohydrate epitopes is greatly reduced or ablated following periodate oxidation (which results in the oxidation of vicinal dihydrol groups; Woodward *et al.* 1985) and antibody binding to some oocyst antigens is periodate-sensitive. For example, binding of mAbs 1B5 (McDonald *et al.* 1995) and OW-IGO (Bonnin *et al.* 1991) is virtually ablated following periodate treatment of the antigens. Conversely, both the recombinant oocyst wall antigen CP41, and its native antigen were periodate insensitive, with neither the apparent molecular mass nor the intensity of mAb binding being affected by periodate oxidation (Jenkins *et al.* 1999).

Lectins are a class of sugar binding proteins of non-immune origin which bind specifically to defined carbohydrate moieties and provide useful probes for investigating the localisation and organisation of carbohydrate containing receptor sites. (Lis and Sharon, 1973). Further information about the nature and localisation of carbohydrate motifs can be obtained from lectin binding studies.

Luft *et al.* (1987) found that sonicated oocyst extracts separated on SDS-PAGE and transferred to nitrocellulose membranes, reacted strongly with various lectins. Concanavalin A (Con A), *Dolichos biflorus* agglutinin (DBA), wheat germ agglutinin (WGA), *Ricinus communis* agglutinin (RCA) and soybean agglutinin (SBA) bound up to 15 bands from *C. parvum* oocyst sonicates, but *Ulex europaeus* agglutinin I (UEA I) and peanut agglutinin (PNA) did not bind. Lectin binding indicated the presence of α-D-mannose (Man) and/or α-D-Glucose (Glc), N-acetyl-α-D-galactosamine, (GalNAc) and N-acetyl-β-D-glucosamine (GlcNAc) and or sialic acid (Neu5Ac) in the blotted proteins.

The complexity of these carbohydrate moieties was further demonstrated in that Con A, DBA and WGA all displayed strong binding to the same eight glycoprotein bands, with apparent molecular masses of 72, 81, 87, 92, 95, 99 & two > 100 kDa. The specificity and avidity of these lectins to these glycoproteins would suggest that the carbohydrate components are relatively complex with α-D-mannosyl and/or α-D-glucosyl and/or N-acetyl-α-D-galactosaminyl terminal residues present, as well as N-acetyl-β-D-glycosaminyl residues and/or N-acetyl-β-glucosamine oligomers (Luft *et al.*, 1987).

In Western blots, antigens of 72, 76, 98 & >100kDa, recognised by immune sera from experimentally infected mice, also bound Con A, indicating their glycoprotein nature (Luft *et al.*, 1987). Treatment of these glycoproteins with mixed glycosidases, (α and β-mannosidases, α and β-galactosidases, α and β-glucosidases, α-L-fucosidase, β-D-xylosidase, α and β-N-acetylglucosaminidases, and α and β-N-acetylgalactosaminidases) led to a significant reduction in antibody binding. Binding of serum antibodies from intraperitoneally injected, but not from orally infected mice, to blotted glycoproteins was reduced by sialic acid specific, neuraminidase treatment, indicating that sialic acid may be an important component of these glycoproteins (Luft, *et al.*, 1987).

*8.3.2 Surface-associated lectin receptors.*Experiments using solubilised antigens from disrupted oocysts provide evidence for the presence of carbohydrate moieties in the compartment analysed, but do not provide information about receptor localisation. Llovo *et al.* (1993) localised carbohydrate residues to *C. parvum* oocyst outer surfaces by agglutinating suspended, intact oocysts with specific lectins. *Codium fragile* agglutinin (CFA), *Codium tomentosum* agglutinin (CTA) and UEA II, which interact with D-GalNAc > D-GlcNAc; D-GlcNAc; and (β-D-GlcNAc)₂ > β-D-GlcNAc, all agglutinated *C. parvum* oocysts at different titres (Table 4). The differences noted in lectin agglutination titre between the 4 *C. parvum* isolates tested may indicate differences in the density of these carbohydrate receptors on the oocyst surface.

Table 4 *Variation in lectin agglutination titre between four C. parvum isolates (from Llovo et al., 1993)*

Lectin	*C. parvum* oocyst isolates			
	C30	C38	C40	C41
Codium fragile	1:256	1:1024	1:1024	1:128
Codium tomentosum	1:128	1:1024	1:64	1:256
Ulex europaeus II	1:64	1:1024	1:1024	1:128

CTA and UEA II (GlcNAc-specific) agglutinated all isolates tested in a saccharide-specific manner: agglutination was ablated by incubating the oocyst / lectin mixtures with GlcNAc. This suggests that GlcNAc residues are present in the outer surface of the oocyst isolates tested. CFA (GalNAc specific) agglutinated *C. parvum* oocysts, whereas other GalNAc specific lectins (DBA, MPA and VVA) did not. Llovo *et al.* (1993) stated that the batch specific data on sugar inhibition provided with the CFA batch used indicated inhibition with both GalNAc and GlcNAc, and the lectin may have been supplied as a mixture of both CFA and CTA. Thus, while only GlcNAc containing residues were specifically identified in the study, GalNAc specific sugars may also be present on *C. parvum* oocysts. No agglutination was observed with WGA and *Datura stramonium* (DSL; D-GlcNAc)₂ specific) or *Lycopersicon esculentum* (LEA), and *Solanum tuberosum* (STA;

D-GlcNAc)₃-specific). The only di-/trisaccharide-specific lectin that agglutinated *C. parvum* oocysts was UEA II. Inhibition assays indicated that CTA-induced agglutination was inhibited by a lower concentration of GlcNAc than UEA-II-induced agglutination, indicating that the monosaccharide-specific lectin displayed a higher binding affinity than the corresponding disaccharide-specific lectin (Llovo, *et al.*, 1993).

Failure to detect lectin agglutination of oocysts does exclude the presence of that specific lectin-binding receptor on the oocyst. The spatial configuration of lectin receptor sites in the intact oocyst may result in monovalent lectin binding, but not agglutination. As such, the presence of other carbohydrate residues cannot be excluded. Although GlcNAc was present on *C. parvum* oocyst surfaces, this receptor cannot be used to distinguish between different *C. parvum* isolates, despite the slight variations observed in lectin agglutinating titre (Llovo, *et al.*, 1993; Table 4).

Nanduri *et al.* (1999) demonstrated the presence of an immunogenic glycocalyx on the outer surfaces of *C. parvum* oocysts, consisting approximately of 82% carbohydrate and 17% protein (2 x 10⁷ oocysts yielded approximately 40 µg of carbohydrate and 8 µg of protein). Glucose was the major sugar (65%) present in the glycocalyx, by carbohydrate compositional analysis, with galactose (12%), mannose, xylose and ribose also present at lower concentrations (4 - 8% of total). Alditol acetate derivative and trimethylsilyl analyses showed that GalNAc was the only amino sugar present in this highly immunogenic glycocalyx. The importance of these carbohydrate moieties to the antigenicity of the glycocalyx was demonstrated by a reduction in antibody binding of approximately 50%, following periodate oxidation.

Studies on lectin binding to the outer surfaces of intact *C. parvum* (5; MD & 4 human isolates), *C. baileyi* (1; Belgium strain, LB 19) and *C. muris* (1; RN 66) oocysts were also conducted at the SPDL (Campbell, 1993). *C. parvum* oocyst purification procedures included filtration (250 µm mesh) and emulsification of faeces, followed by repeated washing in RO water and centrifugation; modified water-ether purification (Bukhari and Smith, 1995); sucrose flotation (Bukhari and Smith, 1995); saturated salt flotation (Weber *et al.*, 1992) and isopycnic Percoll gradient (1.09 g ml⁻¹, 320 mOsm, pH 7.4) centrifugation (Waldman *et al.*, 1986). *C. baileyi* and *C. muris* oocysts were purified by salt followed by sucrose flotation. Lectin binding studies were performed on *C. parvum* and *C. baileyi* oocysts exposed to the following conditions: a) intact oocysts stored in RO water, reacted with fluorescent lectins at both 4°C and 37°C, b) excysted oocysts and c) hypochlorite-treated oocysts, using the following labelled lectins: TRITC-WGA, FITC-PNA, FITC-RCA₁₂₀, FITC-Lima bean agglutinin (FITC-LBL), FITC-SBA, lentil agglutinin (FITC-LCA), FITC-*Bandeireaea simplificola* isolectin B1 (FITC-BS-I), FITC-camels foot tree agglutinin (FITC-BPA), FITC-Con A, and FITC-DBA). Studies on *C. muris* oocysts were limited to condition a). Lectin specificity was demonstrated by preincubation with 0.4M concentrations of the appropriate competing monosaccharides.

a) Lectin binding to intact oocyst walls

Studies on lectin binding to the outer surfaces of intact *C. parvum, C. baileyi* and *C. muris* oocysts were conducted at 4°C and 37°C for 1 h. At 4°C, lectin binding to intact *C. parvum* oocysts was not observed. At 37°C, the pattern of lectin binding to *C. parvum* oocysts was the same, regardless of isolate or purification procedure employed. PNA bound to all *C. parvum* isolates tested in a saccharide-specific manner, but not to *C. baileyi* or *C. muris*, while SBA bound to *C. baileyi* and *C. muris* but not to *C. parvum*. PNA fluorescence was weaker than SBA fluorescence. RCA bound weakly and inconsistently to the outer surface of *C. baileyi* but not to *C. parvum* or *C. muris*. Con A bound to *C. muris* oocysts only (Campbell, 1993; Tables 5 and 6).

Table 5 *Effects of different pretreatments on lectin binding to* Cryptosporidium *oocysts*

	Intact oocysts		Excysted oocyst walls		NaOCl treated oocyst walls	
Lectin	*C. parvum*	*C. baileyi*	*C. parvum*	*C. baileyi*	*C. parvum*	*C. baileyi*
Con A	-	-	+	+	+	+
PNA	+	-	+	+/-	+	-
RCA	-	+	+	+	+	+
SBA	-	+	+	+	+	+
UEA I	-	-	+/-	+/-	-	-
WGA	-	-	+	+	+	+

Key: + = fluorescence observed; +/- = weak and variable fluorescence; - = no fluorescence observed. Data from Campbell (1993).

b) *Lectin binding to the walls of excysted oocysts*
Compared with intact oocysts, more extensive lectin binding patterns occurred using excysted *C. parvum* and *C. baileyi* oocysts. Saccharide-specific staining of both *C. parvum* and *C. baileyi* excysted oocysts was obtained with FITC-Con A, FITC-PNA, FITC-RCA, FITC-SBA and FITC-UEA. TRITC-WGA binding could only be inhibited with 1 mM N,N',N"-Triacetylchitotriose. Con A, RCA and WGA exhibited the same, strong, fluorescence on oocyst walls of both species. PNA bound more intensely to excysted *C. parvum* oocysts than to excysted *C. baileyi* oocysts and SBA bound more intensely to excysted *C. baileyi* oocysts than to excysted *C. parvum* oocysts. The binding of UEA I to excysted oocyst walls was weak and not all the excysted oocyst walls observed bound the lectin (Campbell, 1993; Table 5).

c) *Lectin binding to intact, hypochlorite treated oocyst walls*
Intact, hypochlorite (0.15% for 10 min on ice) treated *C. parvum* and *C. baileyi* oocysts bound FITC-Con A, FITC-RCA and FITC-SBA in a saccharide-specific manner. TRITC-WGA binding could only be inhibited with 1 mM N,N',N"-Triacetylchitotriose. The only differences from excysted oocysts were that oocysts of *C. baileyi* did not bind PNA, and that UEA I was not observed to bind onto either species tested (Campbell, 1993; Table 5).

Surface biotinylation experiments on *C. parvum* (5; MD & 4 human isolates), *C. baileyi* (1; Belgium strain, LB 19) and *C. muris* (1; RN 66) oocysts were also performed at the SPDL (Campbell, 1993). *C. parvum* oocysts were purified as described for the lectin binding studies, above. The presence of surface associated carbohydrates was investigated by labelling intact oocysts with biotin hydrazide (modified from Heizmann and Richards, 1974) following either sodium periodate or galactose oxidase pretreatment. The presence of surface associated primary amines was investigated by labelling intact oocysts with biotin-N-hydroxysuccinimide ester (NHS-biotin, Hurley *et al.*, 1985), and the presence of available disulphide bonds and free sulphydryl groups was investigated by preincubating intact oocysts with dithiothreitol (DTT) or ß-mercaptoethanol (ß-ME) followed by biotin maleimide. Oocysts were biotinylated at 4°C and 37°C and evidence that biotinylated molecules were surface associated in intact oocysts was provided by visualising biotin-bound streptavidin-FITC emissions on oocysts walls by epifluorescence microscopy.

No differences were observed between oocysts biotinylated at either 4°C or 37°C. All *Cryptosporidium* oocyst isolates revealed the same fluorescent pattern irrespective of isolate or purification procedure used. Intact, periodate-treated, but not galactose oxidase-treated, oocysts fluoresced strongly. Fluorescence was predominantly oocyst wall-specific with minimal cytoplasmic staining on ruptured oocysts. Intact, NHS-biotin treated oocysts fluoresced weakly compared with periodate-BHZ treated oocysts. Intact oocysts pretreated with DTT or ß-ME and labelled with biotin maleimide did not fluoresce. Biotin labelling revealed a predominance of carbohydrate groups susceptible to periodate oxidation. Evidence was obtained for the presence of primary amines, but the fluorescence was weak. Neither disulphide bonds (preincubation with DTT or ß-ME followed by biotin maleimide) nor free sulphydryl groups were detected on the outer surface of the intact oocyst isolates studied. Compared to carbohydrates, proteins do not account for a significant proportion of surface exposed residues on *Cryptosporidium* oocysts (Campbell, 1993).

A microphotometric comparison (Leitz, MPV Compact-2) of the relative fluorescent intensities of intact and empty (*in vitro* excysted) *C. parvum* oocyst walls revealed a significant increase in the expression of primary amines (NHS-biotin labelled) in empty oocyst walls, but no increase in fluorescence following periodate-BHZ treatment.

*8.3.3 SDS-PAGE-Western blot analysis of biotinylated C. parvum surfaces.*MD isolate *C. parvum* oocysts of both high (~90% intact) and low (~20% intact) viability were either NHS-biotin or periodate-BHZ labelled, extracted in SDS-PAGE sample buffer and subjected to SDS-PAGE and Western blotting (Campbell, 1993). Blots were developed using an enhanced chemiluminescence system described elsewhere (Colloff *et al.*, 1992; Campbell *et al.*, 1993b). Periodate-BHZ labelling revealed major glycoconjugates of 45 – 55, 70 – 80, 100 - 120 and 200 kDa and also produced more bands than NHS-biotin labelling. More bands were apparent in blots from NHS-biotin-labelled extracts of low viability oocysts than in blots from a high viability population. With low viability oocysts, both the number and profile of NHS-biotin reactive bands were similar to results obtained with periodate-BHZ labelling of intact oocysts. Fewer (5 x 10^5 per lane) oocysts were required to generate optimal intensity Immobilon™ PVDF blot profiles using periodate-BHZ than when using NHS-biotin (10^6 per lane) (Campbell, 1993).

A summary of lectin binding studies is presented in Table 6.

Table 6 *Lectin binding to* Cryptosporidium *oocysts and oocyst antigen preparations*

Lectin	Sugar Specificity	*C. baileyi*	*C. muris*	*C. parvum*	Ref-erence
Allomyrina bichotoma	β-D Gal	ND	ND	-	1
Anguilla anguilla	α-D-Fuc	ND	ND	-	1
Arachis hypogea	D-Gal-β-(1-2)>β-D-GalNH2>α-D-Gal	ND/-	ND/-	-/+	1/2
Bandeireaea simplificola 1	α-D-Gal	+	-	-	2
Bauhinia purpurea	β-Gal(1-3)GalNAc	-	-	-	2

Canavalia ensiformis	α-D-Man>α-D-Man>α-D-GlcNAc	ND	ND	-	1
Codium fragile	D-α-D-Gal>D-GlcNAc	ND/-	ND/-	+/-	1/3
Codium tomentosum	D-GlcNAc	ND	ND	+	1
Concanavalin A	α-D-Man, α-D-Man	+	-	-/+	2/3
Datura stramonium	(D-GlcNAc)2	ND	ND	-	1
Dolichos biflorus	α-D-GalNAc>β-D GalNAc > α-D-Gal	ND/+	ND/-	-/-/+	1/2/3
Lens culinaris	α-D-Man > α-D-Glc	ND/-	ND/-	-/-	1/2
Lima bean agglutinin	GalNAc	+	-	+	2
Limulus polyphemus	NeuNAc	ND/-	ND/-	-/-	1/2
Lycopersicon esculentum	(D-GlcNAc)3	ND	ND	-	1
Maclura promifera	α-D-GalNAc > α-D-Glc	ND	ND	-	1
Pisum sativum	α-D-Man > α-D-Glc > α-D-GlcNAc	ND	ND	-	1
Pilota plumose	α-D-Gal	ND	ND	-	1
Ricinus communis	GalNAc, β-Gal	+	-	-	2
Soy bean agglutinin	GalNAc	+/ND	+/ND	-/+	2/3
Solanum tuberosum	(β-D-GlcNAc)2-5 > β-D-GlcNAc	ND	ND	-	1
Triticum vulgaris	(β-D-GlcNAc)3 >(β-D-GlcNAc)2 >β-D-GlcNAc	ND/ND/-	ND/ND/-	-/-/+	1/2/3
Ulex europaeus I	α-L-Fuc	ND	ND	-	2
Ulex europaeus II	(β-D-GlcNAc)2 >β-D-GlcNAc	ND	ND	+	1
Vicia graminea	(β-D Gal(1-3)GalNAc-α-o) clustered	ND	ND	-	1
Vicia sativa	D-man > D-Glc	ND	ND	-	1
Vicia villosa	D-GalNAc	ND	ND	-	1

Key. ND = no data presented; 1 = Llovo *et al.*, 1993; 2 = Campbell, 1993; 3 = Luft *et al.*,1988.

8.4 Surface charge

Knowledge of the surface charge of *Cryptosporidium* oocysts provides useful information that can used to optimise (oo)cyst removal during coagulation and flocculation.

Measurements of electrophoretic mobility have demonstrated zeta potentials for *C. parvum* oocysts of approximately -25 ± 6 mV at neutral pH (Ongerth and Pecoraro, 1995) and –25 ± 2.8 mV at pH 6 (Drozd and Schwartzbrod, 1996). Surface charge increases slowly with decreasing pH (-35 mV at alkaline pH and reaching zero at pH 2.5 (Drozd and Schwartzbrod, 1996) or pH 4-4.5 (Ongerth and Pecoraro, 1995). Surface charge appears reasonably stable over time, with only a slight decrease in zeta potential from 25 mv to approx. 15 mV over a 1 month period (Drozd and Schwartzbrod, 1996). According to Brush *et al.* (1998) oocyst purification procedures can modify surface charge and these data have been challenged (see Purification procedures, below).

9 BIOPHYSICAL APPROACHES TO OOCYST WALL ANALYSIS

9.1 The *C. parvum* oocyst wall is deformable

When we incubated intact, viable oocysts in dimethyl sulphoxide (10% DMSO) at 37°C for 2 h or sucrose solution (specific gravity 1.18 when cold) at room temperature for up to 1 h, we found that the number of surface folds, demonstrated by FITC-*C*-mAb antibody binding and binding of the surface-reactive, lipophilic, cationic fluorescent dye, octadecyl rhodamine B (R18),increased (Robertson *et al.*, 1993b). Production of folds was reversible when oocysts were then incubated for up to 24 h in RO water, indicating that the oocyst wall was reversibly deformable and that the increase in surface folds was not due to an increase in epitope density. Dependent on the osmolality of the surrounding medium, the oocyst wall of intact, viable oocysts can deform and recover its shape. This may account for anecdotal evidence of ~5 μm diameter oocysts penetrating apertures of <5 μm diameter.

9.2 Absence of lateral diffusion in the *C. parvum* outer oocyst wall

Using the technique of fluorescence recovery after photobleaching (FRAP) for demonstrating lateral diffusion in external surfaces of parasites (Foley *et al.*, 1986; Kennedy *et al.*, 1987; Proudfoot *et al.*, 1990, 1991, 1993a,b), we labelled viable *C. parvum* (MD isolate) oocysts with various fluorogenic probes. In FRAP, the lateral mobility of molecules in the membrane surface is determined by measuring the rate of local changes in the concentration of fluorescent markers.

In practice, a predetermined fluorescent area is quenched with laser light and its fluorescent recolonisation by diffusion from adjacent areas is measured. FRAP provides two measures of diffusion, first, the fraction of the component that is free to diffuse (% recovery, %R) and second, the lateral diffusion coefficient (D_L) of that fraction (Axelrod *et al.*, 1976). Although we demonstrated the insertion of the anionic AF 18 lipophilic probe into intact *C. parvum* oocyst walls, we were unable to demonstrate any lateral movement of AF 18 by FRAP. In addition, in preliminary experiments, no lateral movement of mAbs, carbohydrates or primary amines was detected. However, localised lateral diffusion, beyond the resolution of FRAP, which uses a 1 μm diameter laser beam, cannot be excluded and further investigation into the fluidity of surface-associated molecules is required. As with parasitic nematodes, immobility of *C. parvum* oocyst surface lipid may be a fundamental feature that protects oocysts from the chemical and physical onslaughts of the host and external environment by preventing perturbation of a stable lipid organisation.

9.3 Electrical interrogation of oocyst integrity - dielectrophoresis and electrorotation

The organisation of molecules, macromolecules and organelles within a cell, governs form and function, and the biophysical/biochemical make up of organisms is thought to be unique. Protozoa are regarded as being a simple form of animal life, yet they possess a complex and specialised organisation developed, within a single cell, to cope with their immediate environment. When exposed to electrical fields, the individuality of an organism at the organisational level produces identifiable electrical patterns which can be monitored. The awareness that the integrity of an organism can be interrogated electrically has stimulated renewed interest in the biophysical approaches of dielectrophoresis and electrorotation (ROT). When organisms are exposed to electrical fields, interactions between molecules and the field produce identifiable electrical patterns which can be characterised. Dielectrophoresis and electrorotation, based on analysis using alternating current (A.C.) electrical fields have been used to identify, concentrate and/or assess the viability of *Cryptosporidium* oocysts (Goater and Pethig, 1998).

9.3.1 Dielectrophoresis. Dielectrophoresis describes the motion of neutral particles when exposed to a non-uniform A.C. electric field. Particles become polarised, and the polarisation is dependent upon the interaction between the individual make-up of the particle, the conductivity of the suspending medium and the frequency range applied. The non-uniform electrical field is applied across two electrodes and particles can be attracted to one of the two electrodes or repelled from them. By changing the frequency of the applied voltage, the extent of the migration and the subsequent collection of particles can be altered. Using a defined frequency range, characteristic spectra can be obtained for individual particles.

Discrimination between viable and non-viable *C. parvum* oocysts is dependent upon their dielectrophoretic collection spectra being sufficiently different. The electrical field is applied across two electrodes so that polarised oocysts migrate toward one or other electrode and collect there. The accumulation of oocysts at an electrode is monitored microscopically, and their images captured. Quinn *et al.* (1995) demonstrated that ozone-treated, chlorine-treated and untreated oocysts produced different collection spectra. Oocysts exposed to ozone disinfection (1.5 mg l^{-1} at t_0) demonstrated different collection spectra, collecting to a greater degree at frequencies below 200 kHz and to a lesser degree at 3 - 10 MHz, when compared with their untreated controls.

9.3.2 Electrorotation. Whereas dielectrophoresis can be described as the motion imparted on electrically neutral, but polarised particles, which are subjected to non-uniform electrical fields, electrorotation (ROT) occurs as a result of rotational torque exerted on polarised particles subjected to rotating electrical fields (Goater & Pethig, 1998). Both techniques are dependent upon the relative conductive properties of the particle and the suspending medium. The response of individual oocysts subjected to non-uniform electrical fields is readily observed with ROT. Hence, the possibility of characterising oocysts by measuring relative rotational forces induced by A.C. electrical fields together with the possibility of identifying differences in the A.C. electrodynamics of viable and non-viable organisms exists. Smith *et al.* (1994) and Goater and Pethig (1998) demonstrated that viable oocysts, as determined by their ability to include the vital fluorogenic dye DAPI, to exclude PI and to excyst *in vitro* could be made to rotate clockwise whilst non-viable oocysts, which included DAPI and PI and would not excyst *in vitro* rotated counterclockwise at a pre-determined frequency range.

At lower frequencies, oocyst wall effects are dominant. Applying the pre-determined frequency range at which viable and non-viable oocysts rotate in opposite directions, the effect of changing specific surface-associated properties of oocysts, which affect oocyst viability, can be monitored. As ROT is non-invasive, organisms can be collected and subjected subsequently to a variety of other analytical procedures.

10 EXPERIMENTAL MANIPULATION OF *C. PARVUM* OOCYSTS: CAUSES AND EFFECTS

A massive demand exists for purified *C. parvum* oocysts and most biologists and engineers prefer to work with oocysts which have been purified by a third party! The supply of purified oocysts and their storage can generate a further series of constraints which can influence the biological status of oocysts. Included here are the effects of oocyst purification techniques, age, temperature and sodium hypochlorite treatment.

10.1 Purification Procedures

Oocyst purification methods can affect the nature or antigenicity of molecules present on the oocyst surface. *C. parvum* oocysts washed and suspended in sucrose solution and purified by continuous flow centrifugation followed by sucrose centrifugation, washing and stored in deionised water (DIS) have an electrophoretic mobility of between approximately $+10^{-8}$ and $-10^{-8} m^2 v^{-1} sec^{-1}$, with a linear regression slope of zero over the pH range tested (pH 2-10; Brush *et al.*, 1998). No clear isoelectric point was identified from measurements of DIS purified oocysts but oocysts purified by formalin fixation, ethyl acetate sedimentation followed by Percoll-sucrose centrifugation, washing and storage in deionised water (EAPS) exhibited a strongly linear relationship between electrophoretic mobility and pH, with a pI at pH 2.37 (Brush *et al.*, 1998). Ether or ethyl acetate-based purification methods may remove lipids from the oocyst wall, and exposure to solutions of high ionic strength, such as saturated NaCl flotation and CsCl centrifugation, may affect charge distribution (Brush *et al.*, 1998; Drozd and Schwartzbrod, 1996).

In a collaborative study comparing *in vivo* and *in vitro* surrogates for *C. parvum* viability and infectivity (Clancy *et al.*, 2000), we noticed that viable oocysts purified by sucrose floatation and CsCl centrifugation included DAPI into sporozoite nuclei inconsistently, unlike MD oocysts purified according to Hill *et al.* (1991) or oocysts purified according the methods of Bukhari and Smith (1995) (Table 9). Further analysis indicated that this was not an age dependent phenomenon, but that MD oocysts, exposed to the same concentration of CsCl used for oocyst purification also exhibited reduced and variable DAPI uptake compared to control oocysts. Reducing CsCl concentration reduced this variability. The variability could be overcome by prolonged incubation of CsCl treated MD oocysts in RO water. Subsequent investigations into the effect of oocyst purification procedures on DAPI uptake were reported by Watkins and Smith (1999).

In order to investigate whether the antigenic differences we found in two separate stocks of 'Iowa' isolate oocysts (Ronald *et al.*, 2000) were oocyst dependent or purification procedure dependent, we purified oocysts from a human faecal sample submitted to the SPDL using four different procedures: (a) sucrose flotation followed by CsCl density centrifugation; (b) ether-water centrifugation; (c) sucrose flotation and (d) ether-water centrifugation followed by sucrose flotation. One supplier used procedure (a) and the other procedure (d). SDS-stripped surfaces from purified oocysts were separated by SDS-PAGE and Western blotted. Methods (a) and (d) produced identical antigenic profiles, which were different from the profiles obtained with methods (b) and (c), although methods (b) and (c) also generated identical antigenic profiles (Ronald *et al.*, 2000). Thus, purification procedure can affect profiles of SDS-extracted surface associated antigens as well as oocyst electrophoretic mobility. Our data also indicate that *C. parvum* oocysts of the same 'isolate' are not a "standard" product and that care must be taken when selecting isolates for experimental work, and in extrapolating data obtained from one isolate to another.

MD oocyst suspensions (%DAPI positive > 80%) incubated in saliva at 37°C for between 1 min to 2 h show a significant reduction in excystation observed after a contact time as brief as 1 min and also become less permeable to DAPI (Robertson *et al.*, 1993a). The data on DAPI inclusion suggests some component of saliva causes a temporary reduction in oocyst wall permeability, which is regained following incubation in an acidic solution (e.g. 1 h pre-incubation in 0.1M HCl at room temperature). However saliva pre-treatment of a human isolate of oocysts (%DAPI negative > 85%), purified by water-ether centrifugation and sucrose flotation, did not affect DAPI inclusion. *In vivo*, oocysts whose walls have already been made permeable prior to ingestion, for example by exposure to acidic waters, might be protected from further, possibly lethal, permeablisation in the stomach acid by first becoming impermeable following contact with saliva. Such protection may be due to adsorption of proline-rich salivary proteins.

10.2 Survival of purified oocysts

Robertson *et al.* (1992) studied the survival of different isolates of purified *C. parvum* oocysts under a range of pressures including freezing, desiccation, water treatment processes and also in bovine faeces and various water types, using the fluorogenic vital dyes assay and *in vitro* excystation. Desiccation was lethal to the population studied, but some oocysts withstood exposure to -22°C (90 - 92% killed after 755 h). The water treatment processes investigated did not affect oocyst survival when pH was corrected however, contact with lime, ferric or alum had a significant impact on oocyst survival if pH was not corrected. Oocysts survived well in all water types investigated, including sea water, and when in contact with faeces appeared to develop enhanced impermeability to small molecules which might increase their robustness to some environmental pressures.

Brush *et al.*, (1998) found that the electrophoretic mobility of oocysts stored in deionised water containing antibiotics for between 4 and 121 days does not change with age, although their adhesiveness to polystyrene increased with age. Increasing the ionic strength of the suspending medium influenced the adhesion of recently excreted oocysts more than that of 2 month old oocysts as the former became less adherent to polystyrene.

Surface sterilised, *C. parvum* oocysts, suspended in sterile water at room temperature, disappear in suspension, at varying rates. Decay is genotype dependent, predominantly: the half-life of genotype 2 (n = 4) and a genotype 1 was > 112 days, whereas the majority of genotype 1 oocysts tested (n = 4) had a half-life < 7days (Widmer *et al.*, 2000). The mechanism of disappearance was not investigated but would include lysis or adherence to the container surfaces. Adherence is a possibility, as Brush *et al.* (1998) demonstrated that oocyst adhesiveness to polystyrene increased with age, although the stated differences between genotype 1 and 2 oocysts would have to be reconciled.

10. 3 Temperature induces changes to the *C. parvum* oocyst wall

10.3.1 Selective permeability. Viable C. parvum oocysts are selectively permeable to small molecules, except those that are highly positively charged. We developed a fluorogenic vital dyes assay, based on the inclusion / exclusion of two fluorogens, DAPI (formula weight, 350.2) and PI (formula weight, 668.4) to determine the viability of C. parvum oocysts (Campbell et al., 1992). Two classes of viable oocysts were identified: viable oocysts that included DAPI but excluded PI, and potentially infectious oocysts that did not include either DAPI or PI, had 'viable type' contents by DIC microscopy, and take up DAPI, but not PI, following acid preincubation (Campbell et al., 1992; 1993). In this assay, oocysts are incubated with at 37°C for 2 h, by which time viable oocysts included DAPI into sporozoite nuclei.

During method development, we incubated oocysts and reagents at various temperatures and found that, at temperatures below 37°C, it took longer than 2 h to include DAPI into the nuclei of viable oocysts, although oocysts incubated at 4°C would include DAPI, but not PI, over an extended time period (>12 h). This implies a temperature induced permeability change, effective in viable oocysts only, where the oocyst wall is more permeable to DAPI at 37°C than at 4°C, yet remaining impermeable to PI at both temperatures.

*10.3.2 Lectin reactivity.*Studies on lectin binding to the outer surfaces of intact *C. parvum* oocysts at 37°C for 1 h indicated that PNA bound to all *C. parvum* isolates tested in a saccharide-specific manner, but not to *C. baileyi* or *C. muris* (Campbell, 1993). However, PNA did not bind to intact *C. parvum* oocysts incubated at 4°C. Lectin binding to receptors is temperature-dependent, yet lectins bind at a range of temperatures, and even at lower temperatures equilibrium should be achieved within minutes (Beeley, 1987). PNA binding to intact *C. parvum* oocysts at 37°C, but not at 4°C is suggestive of a temperature-dependent conformational change of receptor molecules expressed on the surface of *C. parvum* oocyst walls.

*10.3.3 C. parvum oocyst wall rupture.*We used the enhanced morphology (DAPI) method of Grimason et al. (1994) to determine the percentage of oocysts which released their nuclei following fracture when exposed to varying numbers of freezing and thawing cycles (Nichols and Smith, 2001, submitted). Various C. parvum oocyst isolates were subjected to freeze thawing by immersion in liquid nitrogen for 1 min and thawing at 65°C for 1 min, then the percentage of ruptured oocysts was enumerated (Grimason et al., 1994). We considered that all oocysts containing fewer than 4 DAPI-positive nuclei were ruptured. When C. parvum oocysts were immersed in liquid nitrogen for 1 min and immediately thawed at 65°C for 1 min, the percentage of ruptured oocysts was dependent upon the isolate and, to a lesser extent, the age of the isolate (Nichols and Smith, 2001, submitted). Iowa isolate oocysts were ruptured more readily than MD or KSU-1 (Kansas State University-1 isolate) oocysts, immaterial of the number of freeze thaw cycles performed (4 freeze-thaw cycles: 93.4 ± 4.3%, 86.2 ± 4.5%, 59.5 ± 6.1%, respectively; 14 freeze-thaw cycles: 95.5 ± 1.3%, 93.1 ± 1.6%, 73.2 ± 2.8%, respectively).

With the KSU-1 isolate, the efficiency of oocyst rupture at 65°C decreased with increasing oocyst age, but not with the Iowa isolate (Table 7), yet at 90°C resistance to rupturing was far more marked with 1 month old Iowa isolate oocysts than with the same oocysts tested 9 months later.

Table 7 *Effect of oocyst age on oocyst rupture after freeze-thawing*

Isolate (age)	% oocysts disrupted	
	Thawing at 65°C	Thawing at 90°C
Iowa (1 month)	99.2	16.5
Iowa (9 months)	99.1	44.8
KSU-1 (5 months)	94.2	49.6
KSU-1 (22 months)	72.1	70.4

10.4 Sodium hypochlorite reduces *C. parvum* surface exposed oocyst epitope reactivity

Immersing *C. parvum* oocysts in sodium hypochlorite solution can lead to epitope loss (Smith *et al.*, 1989; Moore *et al.*, 1998). Moore *et al.* (1998) demonstrated a reduction in epitope reactivity on intact *C. parvum* oocysts stored in 20 and 50 ppm free chlorine for between 4 and 7 days, when reacted with some commercially available FITC-*C*-mAbs and pAbs. Storage reduced the intensities of fluorescence emissions to below detectable limits by flow cytometry, and storage of viable oocysts in 20 and 50 ppm free chlorine reduced oocyst viability to below 50% in 3 days. They argued that oocysts exposed to these levels of free chlorine may loose sufficient surface exposed epitopes recognised by the commercial mAbs and pAbs tested to render them undetectable or, at best, difficult to detect, even though a proportion might remain viable. Surface exposed epitopes recognised by commercially available mAbs and pAbs, were sensitive to free chlorine and periodate oxidation as were SDS-extractable antigens from oocysts on Western blots.

We examined the effects of higher concentrations of NaOCl solution on oocyst epitope expression using the MD isolate and a human isolate pool using a commercially available FITC-*C*-mAb (Northumbria Biologicals, UK) (Smith *et al.*, 1989; Grimason, 1992). Purified MD isolate oocysts were air dried onto microscope slides, immersed in various concentrations of NaOCl (1 - 150,000 mg l^{-1}), and incubated at room temperature for 30 min (Table 8). Only when oocysts were incubated in 1000 mg l^{-1} NaOCl and above was there an identifiable reduction in fluorescence intensity by epifluorescence microscopy. An increase in the number of dull and patchy fluorescent areas was noted at 5,000 mg l^{-1}, while at 15,000 mg l^{-1}, only a weak oocyst wall fluorescence was noted. At 30,000 mg l^{-1} NaOCl and above, fluorescence was ablated (Table 8). When we incubated the pool of purified, human-derived oocysts in suspension on ice for 10 min, 30 min, 60 min or 120 min in 20,000 or 50,000 mg l^{-1} NaOCl, the FITC-*C*-mAb still bound to the intact outer oocyst surfaces, although fluorescence intensity was markedly reduced in all instances, compared with untreated oocysts (Table 8).

Short term exposure of oocysts to NaOCl solutions >1000 mg l^{-1}, either air dried on slides or in suspension, reduced the intensity of bound FITC-*C*-mAb, although sufficient epitope density was present for the considerably reduced fluorescence to be observed after 1 h exposure to 15,000 to 50,000 mg l^{-1} NaOCl.

Incubating *C. parvum* oocysts in 1.75% (17,500 mg l^{-1}) NaOCl for 12 min in an ice bath, results in thinning, perforation or removal of the outer layer of the treated oocyst wall. Occasionally the outer zone of the inner layer may also be removed, but this treatment does not affect the inner zone of the inner layer of the oocyst wall (Reduker *et al.*, 1985b). By extrapolation, we assume that exposure (1h to either 20,000 or 50,000 mg l^{-1} NaOCl (and

Table 8 *Summary of SPDL studies on FITC-C-mAb binding to outer surfaces of C. parvum oocysts incubated in varying concentrations (5 to 50,000 mg l^{-1}) sodium hypochlorite for between 10 min and 4 h at either room temperature or 4°C*

MD isolate oocysts (30 min at RT)	Fluorescence Intensity	Human-derived oocysts (10 - 120 min at 4°C)	Fluorescence Intensity
Control 0 [mg l^{-1}]	+++	Control 0 [mg l^{-1}]	+++
5	+++	20,000; 10 min	+
1000	++	20,000; 60 min	+
5,000	++	20,000; 120 min	+
15,000	+	50,000; 10 min	+
30,000 →150,000	-	50,000; 60 min	+
		50,000; 120 min	+

possibly 15,000 mg l^{-1} NaOCl for 30 min) removed the outer layer of the oocyst. Continued FITC-C-mAb binding indicates that a) the outer wall was not completely removed, b) the epitope is present on the inner layer(s) of the oocyst wall, or c) that a cross-reacting epitope is expressed after NaOCl treatment. Reduced fluorescence on treated oocysts could also be due to NaOCl modification of the FITC-C-mAb epitope or a decrease in its expression density further into the oocyst wall.

For MD oocysts, exposure to 30,000 mg l^{-1} NaOCl and above for 30 min reduced epitope density to a level which could not be visualised using FITC-C-mAb however, sufficient epitope density remained when a pool of human isolate oocysts was exposed to 50,000 mg l^{-1} NaOCl for 60 min, using the same FITC-C-mAb reporter. This is suggestive of MD oocysts possessing a more NaOCl-sensitive epitope or less FITC-C-mAb epitope than the human isolate oocysts tested, possibly with a different arrangement / composition / construction of molecules in the oocyst wall.

In a further study, we labelled viable oocysts with FITC-C-mAb and immediately exposed the labelled, viable oocysts to 1,000 mg l^{-1} and 10,000 mg l^{-1} NaOCl solutions on a microscope slide. Exposure to these concentrations of NaOCl caused the fluorescent, outer oocyst surface to unravel in sheets, within 5 min however, consequent incubation of such NaOCl-treated oocysts in FITC-C-mAb resulted in further FITC-C-mAb labelling.

11 CONCLUSIONS

Previously regarded as a protective covering for sporozoites, current evidence points to a more interactive role for the *Cryptosporidium* oocyst wall. Advances in our understanding of the oocyst wall at the ultrastructural, physiological, biochemical, biophysical, antigenic, molecular and genetic levels have provided us with a better insight into this environmentally robust, protective structure. This multidisciplinary approach has generated valuable data that have been used to identify surface active mechanisms and to predict structures involved in oocyst survival. Not all data contribute to a consolidated understanding of the oocyst wall of viable *C. parvum* oocysts. Some differences in the published data presented may be reconciled by our recent understanding of the genotypic differences within *C. parvum*. Based on known genotypic differences, we might expect phenotypic variation, manifest at the level of the oocyst wall which might explain some anomalies, particularly when using human and animal isolates of *C. parvum*. The potential for incorrectly attributing results obtained using human-derived *C. meleagridis* oocysts to a *C. parvum* oocyst data set also exists and could further complicate the issue.

Cryptosporidium oocyst walls contain lipid moieties, some of which are located in the outer layer of the oocyst wall. These lipids play an important role in determining oocyst wall permeability, surface charge and hydrophobicity, and hence resistance of the oocyst to environmental and possibly disinfection processes. ß-cyclodextrin treatment is selective, as the increased oocyst permeability which culminates in oocyst death, occurs in only a proportion of the oocyst population (Castro-Hermida *et al.*, 2000).

Lumb *et al.* (1988), Tilley *et al.* (1990) and Campbell (1993) report broadly similar banding patterns in *C. parvum* oocyst surfaces, although, again anomalies in the presence or absence of bands accompany these reports. In one study of four human-derived and 1 cervine-derived *C. parvum* isolates tested, 6 common bands were found in all isolates tested, yet a further 4 bands were present in some, but not all isolates tested (Lumb *et al.*, 1988). Some studies indicate that epitopes are conserved between *Cryptosporidium* spp. oocysts, despite differences in their biological behaviour and host specificities (Chrisp *et al.*, 1995). Common antigens of 23 and 32 kDa, as well as antigens in the 40 –180 kDa range have been reported in purified oocyst walls (Lumb *et al.*, 1988), although other studies report a broader spectrum of *C. parvum* oocyst antigens (<14 to >200 kDa). Some of the observed differences may reflect parasite adaptation to different physiological pressures within different host species (Nina, *et al.* 1992b) or the presence of heterogeneous populations of oocysts.

Using a variety of techniques including surface labelling, periodate oxidation and lectin binding, a consensus opinion is that the intact oocyst wall is carbohydrate rich, although polypeptide, primary amine and lipid components have also been demonstrated in oocyst walls. Lectin binding and agglutinating studies also provide strong evidence for the presence of carbohydrate moieties at the outer oocyst interface, with most evidence pointing to the presence of N-acetyl-glucosamine, N-acetyl-galactosamine, mannose and glucose, although some differences are evident in the studies reported (Table 6). Further evidence for the involvement of carbohydrates at this interface can be extrapolated from the fact that many, but not all, of the commercially available FITC-*C*-mAbs are of the IgM isotype, suggestive of a carbohydrate epitope. As commercially available FITC-*C*-mAbs bind to the oocyst with an uniform fluorescence we can assume that the epitope recognised is distributed evenly on the outer oocyst wall. Some *Cryptosporidium* reactive mAbs bind with a patchy granular fluorescence, indicative of the uneven distribution of epitope. Taken together, we can assume that variably distributed epitopes are present on the outer oocyst wall. Dual labelling experiments indicate that the epitopes recognised by two commercially available FITC-*C*-mAbs are similar, rather than being the same and, as neither FITC-*C*-mAb labelled 100% of the population, heterogeneity in epitope expression in an oocyst population may occur. The lectin agglutination titres reported by Llovo *et al.* (1993) also indicate variability in receptor saccharide expression between the 4 *C. parvum* isolates studied, with 5log.$_2$ differences in agglutinating titre demonstrated between isolates.

Biotinylation studies at the SPDL indicate a predominance of surface-exposed carbohydrate moieties (Campbell, 1993). Surface labelling primary amines with NHS-biotin revealed far less reactivity, and while present on outer surfaces of intact oocysts, they are represented to a lesser degree than carbohydrates. Primary amines occur as lysine side-chains in proteins and in polysaccharides as amino sugars (e.g. glucosamine, galactosamine) and the failure to demonstrate disulphide bonds and free sulphydryl groups, accessible to DTT or ß-ME pretreatment and biotin maleimide, suggests that little protein is surface exposed, although the marked primary amine labelling (NHS-biotin) in excysted oocysts indicates that more proteinaceous material is available in the oocyst wall.

Some studies may provide better insight into the biological make up of viable oocysts than others. Experiments using isolated, excysted oocyst walls may not provide an accurate reflection of the constituents of the oocyst wall of viable oocysts, given some of the differences reported between lectin binding and / or surface labelling studies of intact and sonicated or excysted oocyst walls. Compared with intact oocysts, a broader lectin binding profile can be identified using sonicated or excysted oocysts. As periodate oxidation can reduce both lectin and antibody binding, some antigenic analyses using oocyst extracts, sonicates or excysted oocysts may be liable to the same criticisms. Here also, the significance of the differences found in these experiments remains unclear, ranging from variable expression of lectin receptors and antigens to oocyst purification (Brush, *et al.* 1998; Ronald *et al.*, 2000) or preparation artefacts. Specific differences in sample preparation methods (e.g. freeze thaw, sonication, excystation, detergent extraction) could also influence the experimental outcome (Table 9).

Table 9 Cryptosporidium *oocyst purification techniques and storage conditions for the studies presented*

Reference	Oocyst Source	Purification technique and storage conditions	Oocyst treatment
Bonnin *et al.*, 1991.	Human & calf faeces.	Oocysts purified by salt flotation followed by centrifugation in 1.75% NaOCl. Washed and stored in water.	Intact oocysts suspended in sample buffer (65 mM Tris pH 6.8, 1% SDS, 5% sucrose with or without 0.1M dithiothreitol, & subjected to freeze thawing.
Brush *et al.*, 1998.	Calf faeces.	<u>DIS method</u>. Faeces passed through wire mesh, & separated on sucrose gradient in a continuous flow centrifuge. Oocysts stored in water with antibiotics. <u>EAPS method</u>. Faeces passed through wire mesh, oocysts fixed with formaldehyde, sedimented in ethyl acetate, purified by Percoll-sucrose centrifugation, washed and stored in water.	Electrophoretic mobility & hydrophobicity measurements on intact oocysts.
Bukhari & Smith. 1995.	Cervine/ovine (MD) isolate oocysts.	Oocysts purified by :- (a) sucrose flotation. (b) water-ether concentration. (c) zinc sulphate flotation.	*Cryptosporidium*-free faecal samples spiked with oocysts & subjected to different purification techniques. Oocysts subsequently assessed for viability and *in vitro* excystation.

Burkhalter et al., 1998; Schrum et al., 1997; White et al, (1997).	Calf faeces.	Faeces collected into pepsin-HCl solution, stored at 5°C, washed by centrifugation & subjected to continuous flow sucrose centrifugation, followed by CsCl₂ gradient centrifugation.	Lipids extracted in a modified single phase solvent system.
Campbell, 1993.	MD, human & bovine derived oocysts. *C. baileyi* from Dr M. Taylor & *C. muris* from Dr. V. McDonald.	See Hill *et al.* (1991). Human-derived *C. parvum* isolates purified by ether extraction / salt flotation / sucrose flotation or Percoll centrifugation. *C. muris* and *C. baileyi* purified by salt flotation followed by sucrose flotation.	Dual labelling. Intact oocysts labelled in suspension. Western blot analysis. Oocysts boiled in sample loading buffer with or without β-mercaptoethanol for 5 min.
Drozd & Schwartbrod, 1996.	Calf faeces.	Purified by formalin-ethyl acetate sedimentation followed by salt flotation & sucrose gradient centrifugation. Oocysts stored in 2.5% dichromate at 4°C.	Experiments carried out with intact oocysts.
Hill et al., 1991.	Cervine derived oocysts, passaged in sheep (**MD isolate**).	Faeces diluted in water & acidified with HCl (sediments faecal material). Oocyst containing supernatant washed in water & suspended in 1% SDS for 1 h. SDS removed by washing & oocysts stored in HBSS with antibiotics at 4°C.	N/A
Jenkins et al., 1996.	Calf faeces [*C. parvum* (AUCP1) strain].	Filtered through mesh screens, subjected to continuous flow sucrose centrifugation followed by CsCl₂ gradient centrifugation. Stored in water at 4°C.	Protein extracts of oocysts treated with sample buffer containing 2-mercaptoethanol for 3 min in boiling water bath. Analysed by SDS-PAGE and Western blot analysis.
Llovo et al., 1993.	Human faeces.	Filtered through steel mesh & subjected to saccharose gradient centrifugation. Oocysts stored in PBS.	Lectin agglutination of intact oocysts.
Luft et al., 1987.	Calf faeces.	Faeces suspended in PBS, treated with ethyl acetate & subjected to sucrose	Oocyst subjected to sonication & separated into soluble (supernatant) and

		centrifugation. Bacteria removed by filtering through a 3 µm filter in a stirred cell. Oocysts stored in PBS.	insoluble (pellet) fraction by centrifugation. SDS-PAGE: sonicated preparations boiled for 3 min in sample loading buffer with or without β-mercaptoethanol.
Lumb *et al.*, 1988.	Human & goat kid faeces.	Faecal samples mixed with equal volume of 2% potassium dichromate. Freshly prepared sputolysin added to equal volume of faecal sample & allowed to stand for 15 min, ether extracted, washed & subjected to Ficoll gradient centrifugation. Oocysts stored in PBS.	Oocysts excysted in 0.75% trypsin, 2.25% sodium taurocholate in PBS at 37°C. Oocyst walls separated by $CsCl_2$ density centrifugation & soluble antigens prepared by incubating intact oocysts in loading buffer containing 2% SDS and β-mercaptoethanol, heated to 100°C for 5 min.
McDonald *et al.*, 1994.	Calf faeces.	Oocysts purified by ether extraction followed by salt flotation, then washed in 1.5% sodium hypochlorite. Stored in 2.5% dichromate at 4°C.	Oocysts excysted. Oocyst walls & sporozoites centrifuged & resuspended in TBS, then subjected to freeze thawing. Homogenates boiled for 3 min in sample loading buffer with β-mercaptoethanol.
Nanduri *et al.*, 1999.	Human (AIDS) faeces.	Stools filtered through steel mesh, washed & subjected to salt (30%) flotation, washed & filtered through 30 µm nylon membrane. Oocysts treated with 0.05% hypochlorite for 10 min, washed & stored in water with antibiotics at 4 °C.	Oocysts fixed and stained with ruthenium red. Oocysts subjected to periodate oxidation followed by "glycocalyx" removal in 85% phenol. SDS-PAGE and carbohydrate analyses.
Nina *et al.*, 1992a.	Human faeces.	Ether sedimentation & salt flotation.	Oocysts excysted. Oocyst walls & sporozoites centrifuged & resuspended in TBS, then subjected to freeze thawing. Homogenates boiled for 3 min in sample loading buffer with β-mercaptoethanol.
Nina *et al.*, 1992b.	MD isolate oocysts passaged in calves. Other samples	Sucrose or salt flotation.	Oocysts excysted (0.4% bile salts at 37°C for 60-120 min). SDS-PAGE and Western blot analysis. Oocyst antigens prepared by freeze thawing in sample buffer with 1.5 mM

Reference	Source	Purification	Analysis
	purified from faeces. *C. muris* (RN 66) passaged in mice. *C. baileyi* obtained from turkeys.		dithiothreitol. Particulate matter removed by centrifugation & soluble material heated to 100°C for 90 sec in sample buffer & dithiothreitol.
Rannuci *et al.*, 1993.	Calf faeces.	Percoll gradient centrifugation.	Western blot analysis. Oocysts lysed by incubation in Tris-HCl, glycerol and SDS. Proteins in total cell lysates analysed by SDS-PAGE and Western blot analysis.
Robertson *et al.*, 1993.	Calf faeces.	Ether-water extraction, followed by sucrose density centrifugation.	Intact oocysts labelled with FITC-*C*-mAbs and R18. Incubation in DMSO. *In vitro* excysted oocysts.
Ronald *et al.*, 2000.	"Iowa strain" *C. parvum* oocysts passaged in calves.	Sucrose flotation followed by CsCl$_2$ flotation. Ether-water centrifugation. Sucrose flotation. Ether-water centrifugation, followed by sucrose flotation. Oocysts stored in water at 4°C.	Oocyst wall antigens extracted by incubation in 0.5% SDS for 18 h. Analysis by SDS-PAGE & Western blot under non-reducing conditions.
Ronald *et al.*, 2001.	*C. parvum* genotype 1 & 2 oocysts purified from human & animal faeces.	Dilution of sample in water followed by filtration through steel mesh. Sucrose flotation & storage in water.	Oocyst wall antigens extracted by incubation in 0.5% SDS for 18 h. Analysis by SDS-PAGE and Western blot under non-reducing conditions.
Spano *et al.*, 1997.	MD isolate.	See Hill *et al.* (1991).	Immuno-electron microscopy. Oocysts incubated with 2% formaldehyde and 0.8% gluteraldehyde, embedded in gelatine & ultrathin sections prepared.
Tilley *et al.*, 1990.	Oocysts purified from calf faeces & passaged in goats.	Faeces filtered through brass mesh, concentrated by centrifugation & stored in 2.5% K2Cr$_2$O$_7$. Washed in Nonidet P-40, concentrated by sucrose flotation followed	Oocysts subjected to freeze thawing, & solubilised in sample buffer with β-mercaptoethanol.

		by CsCl$_2$ density centrifugation. Stored in PBS at 4°C.	
Weir *et al.*, 2000.	Calf faeces.	Ether water extraction followed by sucrose flotation. Oocysts surface sterilised with ice cold 70% ethanol for 30 min, washed and stored in PBS at 4°C.	Oocyst walls separated from excysted oocysts by IMS. Soluble oocyst extract prepared by boiling oocysts in 0.5% SDS for 1 h.
Widmer *et al.*, 2000.	NEMC1 oocysts obtained from an AIDS patient.	Oocysts isolated by ether extraction followed by isopycnic gradient centrifugation on Nycodenz. Oocysts stored in water.	N/A

Physiological, viability and biophysical studies, based on lipid presence, oocyst deformability, dye permeability, temperature, storage studies, etc. also attract similar criticisms. In addition to epitope and lectin receptor variability, variability of the oocyst wall to fracture following freezing and thawing is isolate, age and thawing temperature dependent. Younger Iowa and KSU-1 isolate oocysts are more susceptible to disruption at 65°C than at 90°C, and younger KSU-1 isolate oocysts more susceptible to disruption at 65°C than older KSU-1 isolate oocysts. However, *C. parvum* oocysts exposed to many of these apparently detrimental processes remain viable. The oocyst survival studies of Robertson *et al.* (1992) using MD isolate oocysts whose purification includes both acid (pH 5-6) and 1% SDS pretreatment indicate that even acid and SDS-modified oocyst walls can protect sporozoites from further adverse insults. Similarly, Fayer (1995) demonstrated that exposure of *C. parvum* oocysts to ≤ 1.31% aqueous NaOCl solution for ≤ 2 h at 21°C significantly reduced, but did not ablate infectivity to BALB/c neonates. While we would not consider exposure to high concentrations of NaOCl a sensible pretreatment for many studies, we must remember that the oocyst wall interface is continuously exposed to various insults both inside and outside the infected host. Apart from extremes of temperature, pH and humidity, some external microenvironments contain bacterial glycosidases and proteases, etc. which can modify the biochemical composition of the oocyst wall (Zuckerman *et al.*, 1997), while water treatment processes can reduce epitope density (e.g. sand, Parker and Smith, 1993; NaOCl, Smith *et al.*, 1989; Moore *et al.*, 1998). Once through this assortment of insults, waterborne oocysts are dried onto microscope slides, deformed, reacted with FITC-*C*-mAbs and recognised by microscopists!

Our preliminary oocyst wall studies with FRAP revealed some similarities with the nematode epicuticle. While the anionic AF 18 lipophilic probe inserted into intact *C. parvum* oocyst walls, no lateral movement could be demonstrated. The nematode epicuticle is also at the environment-parasite interface and must be able to withstand immune and environmental stresses, including cytotoxic assault. As with the adult nematode epicuticle, surface-damaging molecules may be less able to perturb the predominantly stable lipid environment of the *Cryptosporidium* oocyst wall. The survival of nematode parasites in adverse environmental or laboratory induced conditions (including prolonged survival in 1 - 2% SDS) is thought to be due to the conventional membrane being shielded from the

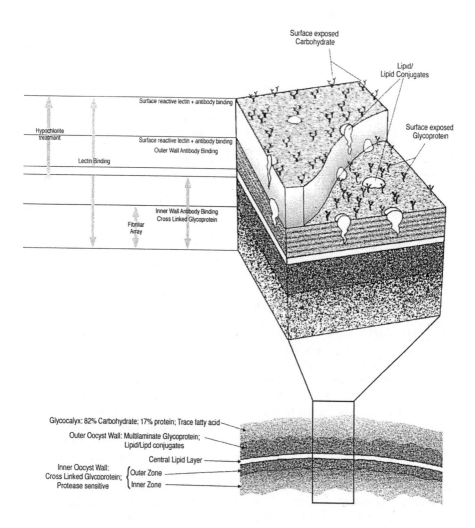

Figure 3 *Model of the* C. parvum *oocyst wall*

environment by glycoconjugates (Proudfoot *et al.*, 1991). Environmental robustness and survival in *Cryptosporidium* oocysts may also be, in part, attributable to a similar mechanism, as survival in SDS (e.g. oocyst purification, Hill *et al.*, 1991; > 12 h at 4°C, our unpublished observations) is also a trait of *Cryptosporidium* oocysts.

By combining data from ultrastructural, biochemical, antigenic, molecular, physiological and biophysical studies, we can develop a simple model of the oocyst wall based on form and function (Figure 3). The oocyst wall consists of outer, central and inner layers, outside of which is a carbohydrate-rich glycocalyx (Nanduri *et al.*, 1999). Oocyst wall functions include protecting sporozoites from various adverse chemical and physical environments, while remaining robust, yet deformable and sensitive to triggers for excystation.

The acidic glycoproteins and lipids in the outer (5 - 10 nm) oocyst wall may contain the immobile AF 18-reactive lipid compartment described and the cholesterol and phospholipids responsible for ß-cyclodextrin induced permeability. Oocyst wall distribution of carbohydrates is more complex. Glucose, galactose and N-acetyl-galactosamine are present in the carbohydrate-rich (82% by composition) glycocalyx (Nanduri *et al.* 1999), with N-acetyl-glucosamine (and possibly N-acetyl-galactosamine) being surface exposed.

Lectin binding studies using intact, excysted and NaOCl treated oocysts indicate that carbohydrates occur deep into the oocyst wall, as do periodate-sensitive mAb immunolocalisation studies. N-acetyl-glucosamine, N-acetyl-galactosamine, mannose and glucose are the most frequently reported lectin receptors. Similarly, biotinylation studies suggest a predominance of surface-exposed carbohydrate moieties. Less common on the intact oocyst surface are primary amines, and the inability to detect available disulphide bonds or free sulphydryl groups suggests that minimal protein is surface exposed, although excysted oocyst walls exhibit a greater carbohydrate and primary amine content.

Variable numbers of proteins and antigens are present in intact oocyst walls. ^{125}I-N-hydroxysuccinimide ester or biotin-N-hydroxysuccinimide ester surface labelling and Western blot analysis of intact oocysts demonstrate that lysine groups of proteins / glycoproteins are available in a surface accessible position, although their number varies, due to either biological or technological differences. Immunolocalisation studies reveal that common antigens are shared between inner and outer oocyst walls, including a native 250 kDa oocyst glycoprotein antigen, present in a fibrillar form in both thick and thin walled oocysts. Its presence in both thick and thin walled oocysts suggests that these immunoreactive molecules are not involved in oocyst robustness. In addition, FITC-*C*-mAb binding to NaOCl stripped *C. parvum* oocysts indicates that the same (or similar) epitope is present in both outer and inner oocyst wall, although its density decreases further into the oocyst wall (possibly the outer zone of the inner layer). Furthermore, unravelling of the FITC-*C*-mAb labelled oocyst wall in sheets with NaOCl suggests a multilaminate structure to the outer oocyst wall.

Epitopes on COWP-190 do not appear to be surface accessible but COWP-190 is abundant in the inner oocyst wall, where the cross-linked native form may exist as a large molecular aggregate, conferring structural stability to the oocyst wall (Spano *et al.*, 1997). The proteinaceous content of the oocyst wall increases with depth. Disulphide bonds appear to be important in antigen structure. Glycosylated, native surface antigens are large, but can be resolved into smaller antigens when disulphide bonds are split.

11.1 A Tough Act to Swallow

Successful propagation of infection depends on the oocyst releasing its sporozoites in the best possible host environment. The deformable viable oocyst of *C. parvum* appears to lack lateral mobility at its surface, but is selectively permeable to small molecules in a temperature-dependent manner. Lectin binding is also temperature dependent. The saccharide-rich oocyst surface masks polypeptides and lipids present below it. This is a function of viable oocysts, as dead oocysts have more surface-exposed polypeptides. Temperature-induced increases in DAPI permeability and lectin receptor expression as well as the conversion of DAPI impermeable oocysts to DAPI permeable oocysts following acid incubation suggest that these two triggers for excystation increase oocyst wall permeability allowing small aqueous molecules to enter viable oocysts. Whether increased permeability is associated with, or due to, any conformational change brought about by lectin receptor exposure, *in vivo*, is not known. On ingestion, viable oocysts with permeable walls are protected from further, possibly lethal, permeablisation in stomach acid by becoming impermeable following contact with saliva. Saliva can prevent premature permeability (Robertson *et al.*, 1993a) and possibly dissolution of the oocyst wall (Fayer and Leek, 1984) which will reduce the likelihood of excystation *in vivo* occurring before oocysts reached the small intestine, the best possible host environment. Increase in turgor pressure within the oocyst together with the dissolution of the oocyst wall at the site of the suture, causing release of the sporozoites, completes the process of excystation. While exogenous proteases do not affect the dynamics of *in vitro* excystation, trypsin enhances sporozoite motility, and possibly invasion of host cells.

C. parvum oocysts afford a stable and resistant covering for infective sporozoites, providing protection from both host and external influences, yet remaining sensitive to triggers known to initiate excystation. While we have some insight into the mechanisms by which the oocyst wall provides these functions, the molecular structures and topographies controlling form, function and disinfection sensitivity have still to be unravelled.

ACKNOWLEDGEMENTS

We acknowledge the contributions of A.T. Campbell, L.J. Robertson, R.A.B. Nichols, A.M. Grimason and J.F.W. Parker to the studies performed at the SPDL and thank Professor J.R. Kusel, Division of Biochemistry and Molecular Biology, University of Glasgow for constructive comments and critical review of the manuscript. We thank the UK Department of the Environment (now Department of the Environment, Transport and the Regions) the Drinking Water Inspectorate, Water Research centre, Ministry of Agriculture, Fisheries and Food and the Overseas Development Administration, Health and Population Division (now Department for International Development) for financial support.

12 REFERENCES

Anonymous (1980). Legislation. *Official Journal of the European Communities* L299 **23,** 11-29.
Anonymous (1990). Isolation and identification of *Giardia* cysts, *Cryptosporidium* oocysts and free living pathogenic amoebae in water etc. 1989. Methods for the examination of waters and associated materials. London, HMSO. 30pp.
Anonymous (1995a). Cryptosporidium in water supplies. Second Report of the Group of Experts; Chairman, Sir John Badenoch. Department of the Environment, Department of Health. London, UK. HMSO. 108pp.
Anonymous (1995b). Proposal for a Council Directive concerning the quality of water intended for human consumption. *Commission of the European Communities*, 1995 99 pp.
Anonymous (1999a). Isolation and identification of *Cryptosporidium* oocysts and *Giardia* cysts in waters 1999. Methods for the examination of waters and associated materials. London, HMSO. 44pp.

Anonymous (1999b). UK Statutory Instruments No. 1524. The Water Supply (Water Quality) (Amendment) Regulations 1999. The Stationery Office, Ltd. 5pp.

Axelrod, D., Koppel, D.E., Schlessinger, J., Elson, E. and Webb, W.W. (1976). Mobiliy measurement by analysis of fluorescence photobleaching recovery kinetic. *Biophys. J.* **16**, 1055-1069.

Beeley, J.G. (1987). Glycoproteins and proteoglycans. In: *Laboratory Techniques in Biochemistry and Molecular Biology* (Burdon, R.H. and van Knippenberg, P.H. editors), Elsevier, Amsterdam, The Netherlands. **16**, 5-28.

Bjorneby, J.M.; Riggs, M.W. and Perryman, L.E. (1990). *Cryptosporidium parvum* merozoites share neutralisation sensitive epitopes with sporozoites. *J. Immunol.* **145**, 298-304.

Bonnin, A., Dubremetz, J.F. and Camerlynck, P. (1991). Characterisation and immunolocalization of an oocyst wall antigen of *Cryptosporidium parvum* (Protozoa: Apicomplexa). *Parasitology* **103**, 171-177.

Brush, C.F.; Walter, M.F.; Anguish, L.J. and Ghiorse, W.C. (1998). Influence of pre-treatment and experimental conditions on electrophoretic mobility and hydrophobicity of *Cryptosporidium parvum* oocysts. *Appl. Environ. Microbiol.* **64**, 4439-4445.

Bukhari, Z. and Smith, H.V. (1995). Effect of three concentration techniques on viability of *Cryptosporidium parvum* oocysts recovered from bovine faeces. *J. Clin. Microbiol.* **33**, 2592-2595.

Burkhalter, R.S., Smith, C.A., White D.C., Fayer, R. and White, A.B. (1998). The signature 10-hydroxy stearic acid thought to correlate with infectivity in oocysts of *Cryptosporidium* species is an artefact. *Lipids.* **33**, 829-833.

Campbell, A.T., Robertson, L.J. and Smith, H.V. (1992). Viability of *Cryptosporidium parvum* oocysts: Correlation of in vitro excystation with inclusion or exclusion of fluorogenic vital dyes. *Appl. Environ. Microbiol.* **58**, 3488-3493.

Campbell, A.T. (1993). *Cryptosporidium* oocysts: detection, viability and biochemical analysis of oocyst walls. PhD thesis, University of Strathclyde, Glasgow. 105pp.

Campbell, A.T., Smith, H.V. and Stimson, W.H. (1993a). Production of high affinity, species specific monoclonal antibodies against oocysts of *Cryptosporidium parvum*. Report for contract no. 09006/14789 Water Research centre (WRc), Marlow, Bucks, UK.

Campbell, A.T., Robertson, L.J. and Smith, H.V. (1993b). Detection of oocysts of *Cryptosporidium* by enhanced chemiluminescence. *J. Microbiol. Methods* **17**, 297-303.

Castro-Hermida, J.A., Friere-Santos, F., Otieza-Lopez, A.M. and Ares-Mazas, E. (2000) Unexpected activity of beta cyclodextrin against experimental infection by *Cryptosporidium parvum. J. Parasitol.* **86**, 1118-1120.

Chrisp, C.E., Suckow, M.A., Fayer, R., Arrowood, M.J., Healey, M.C. and Sterling, C.R. (1991). Comparison of the host ranges and antigenicity of *Cryptosporidium parvum* and *Cryptosporidium wrairi* from Guinea pigs. *J. Protozool.* **39**, 406-409.

Chrisp, C.E., Mason, P. and Perryman, L.E. (1995). Comparison of *Cryptosporidium parvum* and *Cryptosporidium wrairi* by reactivity with monoclonal antibodies and ability to infect severe immunodeficient mice. *Infect. Immun.* **63**, 360-362.

Clancy, J.L., Bukhari, Z., McCuin, R.M. and Clancy, T.P., Marshall, M.M., Korich, D.G., Fricker, C.R., Sykes, N., Smith, H.V., O'Grady, J.E., Rosen, J.P., Sobrinho, J. and Schaefer III, F.W. (2000). *Cryptosporidium* viability and infectivity methods. AWWA research foundation and English Department of the Environment Transport and the Regions, Drinking Water Inspectorate. 137pp.

Colloff, M.J., Howe, C.W., McSharry, C. and Smith, H.V. (1992). Characterisation of IgE antibody binding profiles of sera from patients with atopic dermatitis to allergens of the domestic mites *Dermatophagoides pteronyssinus* and *Euroglyphus maynei*, using enhanced chemiluminescent immunoblotting. *Int. Arch. Allergy. Appl. Immunol.* **97**, 44-49.

Croll, B and Hall, T. (1997). Control of *Cryptosporidium* during drinking water treatment - technological options. In: *Cryptosporidium in water - the challenge to policy makers and water managers.* Chartered Institution of Water and Environmental Management Symposium. Glagsow, 4th December, 1997. pp. 43-57.

Current, W.L. and Garcia, L.S. (1991). Cryptosporidiosis. *Microbiol. Reviews.* **4**, 325-358.

Current, W.L. and Reese, N.C. (1986). A comparison of endogenous development of three isolates of *Cryptosporidium* in suckling mice. *J. Protozool.* **33**, 98-108.

Current, W.L., Upton, S.J. and Haynes, T.B. (1986). The life-cycle of *Cryptosporidium baileyi* sp. (Apicomplexa: Cryptosporidiidae) infecting chickens. *J. Protozool.* **33**, 289-296.

Drozd, C. and Schwartzbrod, J. (1996). Hydrophobic and cell surface properties of *Cryptosporidium parvum. Appl. Environ. Microbiol.* **63**, 1227-1232.

Drury, D. (2001). *Cryptosporidium* Regulations - The Overall Pattern of Results. In: *Regulatory Cryptosporidium: 11 months on.* (Hoyle, B., editor) Royal Society of Chemistry, Institution of Water Officers and Society of Chemical Industry. Paper 5-2.

Fayer, R. and Leek, R.G. (1984). The effect of reducing conditions, medium, pH, temperature and time on *in vitro* excystation of *Cryptosporidium. J. Protozool.* **31**, 567-569.

Fayer, R.; Speer, C.A. and Dubey, J.P. (1990). General biology of *Cryptosporidium.* In: *Cryptosporidiosis of man and animals.* (Dubey, J.P, Fayer, R.and Speer, C.A., editors). CRC Press, Boca Raton Fl. USA. pp. 1 – 29.

Foley, M., MacGregor, A.N., Kusel, J.R., Garland, P.B., Downie, T. and Moore, I. (1986). The lateral diffusion of lipid probes in the surface membrane of *Schistosoma mansoni. J. Cell. Biol.* **103**, 807-818.

Goater, A.D. and Pethig, R. (1998). Electrorotation and dielectrophoresis. *Parasitology.* **117(Suppl.),** S177-S189.

Grimason, A.M. (1992). The occurrence and removal of *Cryptosporidium* sp. oocysts and *Giardia* sp. cysts in surface, potable and wastewater. PhD thesis, University of Strathclyde, Glasgow. 238 pp.

Grimason, A.M., Smith, H.V., Parker, J.F.W., Bukhari, Z., Campbell, A.T. and Robertson, L.J. (1994). Application of DAPI and immunofluorescence for enhanced identification of *Cryptosporidium* spp oocysts in water samples. *Water Res.* **28**, 733-736.

Hall, T. and Pressdee, J.R. (1995). Removal of *Cryptosporidium* during water treatment. In: *Proceedings of workshop on treatment optimisation for Cryptosporidium removal from water supplies.* (West, P.A. and Smith, M.S., editors). HMSO. pp. 25-31.

Harris, J.R. and Petry, F. (1999). *Cryptosporidium parvum:* Structural components of the oocyst wall. *J. Parasitol.* **85**, 839-349.

Heizmann, H. and. Richards, F.M. (1974). The avidin-biotin complex for specific staining of biological membranes in electron microscopy. *Proc. Natl. Acad. Sci. USA.* **71**, 3537-3541.

Hill, B.D., Blewett, C.R., Dawson, A.M. and Wright, S. (1991). *Cryptosporidium parvum:* investigation of sporozoite excystation *in vivo* and association of merozoites with intestinal mucus. *Res. Vet. Sci.* **51**, 264-267.

Hurley, W.L., Finkelstein, E. and Holst, B.D. (1985). Identification of surface proteins on bovine leukocytes by a biotin-avidin protein blotting technique. *J. Immunol. Methods* **85**, 195-198.

Iseki, M., Maekawa, T., Moriya, K., Uni, S. and Takada, S. (1989). Infectivity of *Cryptosporidium muris* (Strain RN66) in various laboratory animals. *Parasitol. Res.* **75**, 218-222.

Jenkins, M.C., Trout, J., Murphy, C., Harp, J.A., Higgins, J., Wergin, W. and Fayer, R. (1999). Cloning and expression of a DNA sequence encoding a 41 Kilodalton *Cryptosporidium parvum* oocyst wall protein. *Clin. Diag. Lab. Imm.* **6**, 912-920.

Karim, M.J., Basak, S.C. and Trees, A.J. (1996). Characterisation and immunoprotective properties of a monoclonal antibody against the major oocyst wall protein of *Eimeria tenella*. *Infect. Immun.* **64**, 1227-1232.

Kennedy, M.W., Foley, M., Knox, K., Harnett, W., Worms, M.J., Kusel, J.R., Birmingham, J. and Garland, P.B. (1987). Are the biophysical properties of the surface lipid of filariae different from other parasitic nematodes. In *Molecular paradigms for eradicating Helminthic parasites*. (MacInnes, A.J., editor), Alan R. Liss, New York. pp. 289-300.

Levine, N.D. (1973). Introduction, history and taxonomy. In: *The Coccidia* (Hammond, D.M. and Long, P.L. editors), University Park Press, Baltimore, Maryland, USA. pp.1-22.

Lis, H. and Sharon, N. (1973). The biochemistry of plant lectins. *Annu. Rev. Biochem.* **42**, 541-574.

Llovo, J., Lopez, A., Fabregas, J. and Munoz, A. (1993). Interaction of lectins with *Cryptosporidium parvum*. *J. Infect. Dis.* **167**, 1477-1480.

Lorenzo, M.J., Ben, B., Mendez, E., Villacorta, I. and Ares-Mazas, M. (1993). *Cryptosporidium parvum* oocyst antigens recognised by sera from infected asymptomatic adult cattle. *Vet. Parasitol.* **60**, 17-25.

Luft, B.J. Payne, D. Woodmansee, D. and Kim C.W. (1987). Characterisation of *Cryptosporidium* antigens from sporulated oocysts of *Cryptosporidium parvum*. *Infect. Immun.* **55**, 2436-2441.

Lumb, R., Lanser, J.A. and O'Donoghue, P.J. (1988). Electrophoretic and immunoblot analysis of *Cryptosporidium* oocysts. *Immunol. Cell. Biol.* **66**, 369-370.

Lumb, R., Smith, P.S., Davies, R., O'Donoghue, P.J., Atkinson, H.M. and Lanser, J.A. (1989). Localization of a 23,000 MW antigen of *Cryptosporidium parvum* by immunoelectron microscopy. *Immunol. Cell. Biol.* **67**, 267-270.

McClellan, P. (1998). Sydney Water Inquiry. Third Report. Assessment of the contamination events and future directions for the management of the catchment. http://www.premiers.nsw.gov.au/publications/pubs_dload/sydwater.../rep3ch1.ht

McDonald, V., Deer, R.M.A., Nina, J.M.S., Wright, S., Chiodini, P.L. and McAdam K.P.W.J. (1991). Characteristics and specificity of hybridoma antibodies against oocyst antigens of *Cryptosporidium parvum* from man. *Parasite Immunol.* **13**, 251-259.

McDonald, V., McCrossan, M.V. and Petry, F. (1995). Localisation of parasite antigens in *Cryptosporidium parvum* infected epithelial cells using monoclonal antibodies. *Parasitology*. **110**, 259-268.

Mitschler, R.R., Welti, R. and Upton, S.J. (1994). A comparative study of the lipid composition of *Cryptosporidium parvum* (Apicomplexa) and Madin-Daby bovine kidney cells. *J. Eukaryot. Microbiol.* **41**, 8-12.

Morgan, U.; Weber, R.; Xiao, L.; Sulaiman, I.; Thomson, R.C.A.; Ndritu, W.; Lal, A.; Moore, A. and Depluzes, P. (2000). Molecular characterisation of *Cryptosporidium* isolates obtained from human immunodeficiency virus infected individuals living in Switzerland, Kenya and the United States. *J. Clin. Microbiol.* **38,** 1180-1183.

Moore, A.G., Vesey, G., Champion, A., Scandizzo, P., Deere, D., Veal, D. and Williams, K.L. (1998). Viable *Cryptosporidium parvum* oocysts exposed to chlorine or other oxidising conditions may lack identifying epitopes. *Int. J. Parasitol.* **28,** 1205-1212.

Nanduri J., Williams S., Aji T. and Flanigan TP. (1999). Characterization of an immunogenic glycocalyx on the surfaces of *Cryptosporidium parvum* oocysts and sporozoites. *Infect. Immun.* **67,** 2022-2024.

Nina, J.M.S., McDonald, V., Dyson, D.A., Catchpole, J., Uni, S., Iseki, M.I., Chiodini, P.L. and McAdam, P.W.J. (1992a). Analysis of oocyst wall and sporozoite antigens from three *Cryptosporidium* species. *Infect. Immun.* **60,** 1509-1513.

Nina, J.M.S., McDonald, V., Deer, R.M.A., Wright, S.E., Dyson, D.A., Chiodini. P.L. and McAdam, K.P.W.J. (1992b). Comparative study of the antigenic composition of oocyst isolates of *Cryptosporidium parvum* from different hosts. *Parasite Immunol.* **14,** 227-232.

Ohtani, Y., Irie, T., Uekema, K., Fukunaga, K. and Pitha, J. (1989). Differential effects of alpha-, beta--3- and gamma- cyclodextrins on human erythrocytes. *Euro. J. Biochem.* **186,** 17-22.

Ongerth, J.E. and Pecoraro, J.P. (1996). Technical report: Electrophoretic mobility of *Cryptosporidium* oocysts and Giardia cysts. *J. Env. Eng.* **122,** 228-231.

Ortega-Mora, L.M.; Tronsco, J.M.; Rojo-Vazquez, F.A. and Gonzalez-Bautista, M. (1992). Cross reactivity of polyclonal serum antibodies generated against *Cryptosporidium parvum* oocysts. *Infect. Immun.* **60,** 3442-3445.

Parker, J.F.W. and Smith, H.V. (1993). Destruction of *Cryptosporidium* oocysts by sand and chlorine. *Water Res.* **27,** 729-731.

Pedraza-Diaz, S., Amar, C. and McLauchlin, J. (2000). The identification and characterisation of an unusual genotype of *Cryptosporidium* from human faeces as *Cryptosporidium meleagridis. FEMS Microbiol. Lett.* **189,** 189-194.

Pedraza-Diaz, S., Amar, C., Iversen, A.M., Stanley, P.J. and McLauchlin, J. (2001). Unusual *Cryptosporidium* species recovered from human faeces: first description of *Cryptosporidium felis* and *Cryptosporidium* "dog-type" from patients in England. *J. Med. Microbiol.* **50,** 293-296.

Proudfoot, L., Kennedy, M.W., Kusel, J.R. and Smith, H.V. (1991). Biophysical characterisation of nematode surfaces. *In: Antigens, Membranes and Genes* (Kennedy, M.W., editor), Taylor and Francis, U.K. pp. 1-26.

Proudfoot, L., Kusel, J.R., Smith, H.V., Harnett, W., Worms, M.J. and Kennedy, M.W. (1993a). Rapid changes in the surface of parasitic nematodes during transition from pre- to post-parasitic forms. *Parasitology* **107,** 107-117.

Proudfoot, L., Kusel, J.R., Smith, H.V. and Kennedy, M.W. (1993b). External stimuli and intracellular signalling in the modification of the nematode surface during transition to the mammalian host environment. *Parasitology* **107,** 559-566.

Quinn, C.M., Archer, G.P., Betts, W.B. and O'Neill, J.G. (1995). An image analysis enhanced rapid dielectrophoretic assessment of *Cryptosporidium parvum* oocyst treatment. In: *Protozoan Parasites and Water.* (Betts, W.B., Casemore, D.P., Fricker, C.R. Smith, H.V. and Watkins, J., editors) Royal Society of Chemistry, Cambridge, UK. pp. 125-132.

Ranucci, L., Muller, H.M., La Rosa, G., Reckman, I., Morales, M.A.G., Spano, F. and Crisanti, A. (1993). Characterization and immuno localization of a *Cryptosporidium* protein containing repeat amino acid motifs. *Infect. Immun.* **61,** 2347-2356.

Reduker, D.W., Speer, C.A. and Blixt, J.A. (1985a). Ultrastructure of *Cryptosporidium parvum* oocysts and excysting sporozoites as revealed by high resolution scanning electron microscopy. *J. Protozool.* **32,** 708-711.

Reducker, D.W., Speer, C.A. and Blixt J.A. (1985b). Ultrastructural changes in the oocyst wall during excystation of *Cryptosporidium parvum* (Apicomplexa: Eucoccidiorida). *Can. J. Zool.* **63,** 1892-1896.

Robertson, L.J., Campbell, A.T. and Smith, H.V. (1992). Survival of oocysts of Cryptosporidium parvum under various environmental pressures. *Appl. Environ. Microbiol.* **58,** 3494-3500.

Robertson, L.J., Campbell, A.T. and Smith, H.V. (1993a). *In vitro* excystation of *Cryptosporidium parvum. Parasitology* **106,** 13-29.

Robertson, L.J., Campbell, A.T. and Smith, H.V. (1993b). Induction of folds or sutures on the walls of *Cryptosporidium parvum* oocysts and their importance as a diagnostic feature. *Appl. Environ. Microbiol.* **59,** 2638-2641.

Ronald, A., O'Grady, J.E. and Smith, H.V. 2001. (2000). Antigenic differences in *Cryptosporidium parvum* oocysts. The "Iowa isolate" enigma. *Royal Society of Tropical Medicine and Hygiene in Scotland. Meeting at the University of Glasgow, Glasgow, 17 May 2000. In: Trans. R. Soc. Trop. Med. Hyg.* **94,** 616-619.

Ronald, A.; O'Grady, J.E. and Smith, H.V. (2001). Antigenic analysis of *Cryptosporidium parvum* isolates of human and animal origin. Submitted.

Rose, J.B., Landeen, L.K., Riley, K.R. and Gerba, C.P. (1989). Evaluation of immunofluorescence techniques for detection of *Cryptosporidium* oocysts and *Giardia* cysts from environmental samples. *Appl. Environ. Microbiol.* **55,** 3189-3196.

Schrum, D.P., Alugupalli, S., Kelly, S.T., White, D.C. and Fayer, R. (1997). Structural characterisation of a "signature" Phosphatidylethanolamine as the major 10-Hydroxy stearic acid containing lipid of *Cryptosporidium parvum* oocysts. *Lipids* **32,** 789-793.

Slifko, T.R.; Coulliette, A.; Huffman, D.E. and Rose, J.B. (2000). Impact of purification procedures on the viability and infectivity of *Cryptosporidium parvum* oocysts. *Water Sci. & Technol.* **41,** 23-29.

Smith, H.V., McDiarmid, A., Smith, A.L., Hinson, A.R. and Gilmour, R.A. (1989a). An analysis of staining methods for the detection of *Cryptosporidium* spp. oocysts in water-related samples. *Parasitology* **99,** 323-327.

Smith, H.V., Smith, A.L. and Girdwood, R.W.A. (1989b). The effect of free chlorine on the viability of *C. parv*um oocysts. Publication number PRU 2023-M. Water Research centre (WRc), Marlow, Bucks, UK.

Smith, H.V. and Rose, J.B. (1990). Waterborne *Cryptosporidiosis. Parasitol. Today* **6,** 8-12.

Smith, H.V., Burt, J.P.H., Bukhari, Z., Pethig, R. and Parton, A. (1994). Alternating current electrical fields and *Cryptosporidium parvum* oocyst viability. *Trans. R. Soc. Trop. Med. Hyg.* **88,** 499-506.

Smith, H.V. (1995). Emerging technologies for the detection of protozoan parasites in water. In: *Protozoan Parasites and Water.* (Betts, W. B., Casemore, D., Fricker, C., Smith, H. V. & Watkins, J., editors), The Royal Society of Chemistry, Cambridge, UK. pp. 108-114.

Smith, H.V. (1996). Detection of *Cryptosporidium* and *Giardia* in water. In: *Molecular approaches to environmental microbiology.* (Pickup, R.W. & Saunders, J.R., editors), Ellis-Horwood Ltd., Hemel Hempstead, UK. Chapter 10, 195-225.

Smith, H.V. (1998). Detection of parasites in the environment. *Parasitology* **117(Suppl.)**, S113-S141.

Smith, H.V. and Rose, J.B. (1998). Waterborne cryptosporidiosis: Current status. *Parasitol. Today* **14**, 14-22.

Smith, H.V. and Nichols, R.A.B. (2001). Case study of health effects of *Cryptosporidium* in drinking water. Article 4.12.4.8. *UNESCO-EOLSS Encyclopaedia of Life Support Systems* –Theme - Environmental Toxicology and Human Health, In press.

Spano, F., Puri, C., Ranucci, L., Putignani, L. and Crisanti, A. (1997). Cloning of the entire COWP gene of *Cryptosporidium parvum* and ultrastructural localization of the protein during sexual parasite development. *Parasitology* **114**, 427-437.

Stotish, R.L., Wang, C.C. and Meyenhoeffer, M. (1978). Structure and composition of the oocyst wall of *Eimeria tenella*. *Parasitology* **64**, 1074-1081.

Tilley, M., Upton, S.J., Blagburn, B.L. and Anderson, B.C. (1990). Identification of outer oocyst wall proteins of three *Cryptosporidium* (Apicomplexa: Cryptosporidiidae) species by ^{125}I surface labelling. *Infect. Immun.* **58**, 252-253.

Tyzzer, E.E. (1912). *Cryptosporidium parvum* (sp. nov.), a coccidium found in the small intestine of the common mouse. *Arch. Protistenkd.* **26**, 394-412.

Tzipori, S. and Griffiths, J.K. (1998). Natural history and biology of *Cryptosporidium parvum*. *Adv. Prasitol.* **40**, 5-36.

Tzipori, S. (1988). Cryptosporidiosis in Perspective. *Adv. Parasitol.* **27**, 63-129.

United States Environmental Protection Agency (1989). National primary drinking water regulations; filtration and disinfection; turbidity; *Giardia lamblia*, viruses, *Legionella*, and heterotrophic bacteria. *Fed. Reg.* **54**, 27486-27541

United States Environmental Protection Agency (1994). National Primary Drinking Water Regulations: Enhanced surface water treatment requirements; proposed rule. *Fed. Reg.* **59**, 38832-38858.

Upton, S.J. and Current, W.L. (1985). The species of *Cryptosporidium* (Apicomplexa: Cryptosporidiidae) infecting mammals. *J. Parasitol.* **71**, 625-629.

Waldman, E., Tzipoi, S. and Forsyth, J.R.L. (1986). Separation of *Cryptosporidium* species oocysts from faeces by using a Percoll density gradient. *J. Clin. Microbiol.* **23**, 199-200.

Upton, S.J. and W.L. Current. (1985). The species of *Cryptosporidium* (Apicomplexa: Cryptosporidiidae) infecting mammals. *J. Parasitology.* **71**, 625-629.

Ward, H. and Cevallos, A.M. 1998. *Cryptosporidium:* molecular basis of host parasite interactions. *Adv. Parasitol.* **40**, 151-178.

Watkins, J. and Smith, M. (1999). Viability assessment of *Cryptosporidium parvum* oocysts by staining with vital dyes. In: *Proceedings of the Conference on Isolation, Propagation and Characterisation of Cryptosporidium.* (Gasser, R.B. and O'Donoghue P., editors) *Int. J. Parasitol.* **29**, 1384-1386.

Weber, R., Bryan, R.T. and Juranek, D.D. (1992). Improved stool concentration procedure for the detection of *Cryptosporidium* oocysts in fecal specimens. *J. Clin. Microbiol.* **30**, 2869-2873.

Weir, C., Vesey, G., Slade, M., Ferrari, B., Veal, D.A. and Williams, K. (2000). An Immunoglobulin G1 antibody highly specific to the wall of *Cryptosporidium* oocysts. *Clin. Diagn. Lab. Immunol.* **7**, 745-750.

White, D.C., Alugupalli, S., Schrum, D.P., Kelly, S.T., Sikka, M.K. Fayer, R. and Kaneshiro, E.S. (1997). Sensitive quantitative detection / identification of infectious *Cryptosporidium parvum* oocysts by signature lipid biomarker analysis. In: *International Symposium on Waterborne Cryptosporidium Proceedings, Newport Beach, California.*

(Fricker, C.R., Clancy, J.C. and Rochelle, P.A., editors) American Water Works Association, Colorado, USA. pp. 53-60.

Widmer, G., Akiyoshi, D., Buckholt, M.A., Feng, X., Rich, S.M., Deary, K.M., Bowman, C.A., Xu, P., Wang, Y., Wang, X., Buck, G.A. and Tzipori, S. (2000). Animal propogation and genomic survey of a genotype 1 isolate of *Cryptosporidium parvum*. *Mol.. Biochem. Parasitol.* **108,** 187-190.

Woodward, M.P., Young, W.W. and Bloodgood, R.A. (1985). Detection of monoclonal antibodies specific for carbohydrate epitopes using periodate oxidation. *J. Immunol. Meth.* **78,** 143-153.

Zuckerman U., Gold, D., Shelef, G., Yuditsky, A., and Armon, R. (1997). Microbial degradation of *Cryptosporidium parvum* by *Serratia marcescens* with high chitinolytic activity. In: *International Symposium on Waterborne Cryptosporidium Proceedings, Newport Beach, California.* (Fricker, C.R., Clancy, J.C. and Rochelle, P.A., editors) American Water Works Association, Colorado, USA. pp. 297-304.

MOLECULAR EPIDEMIOLOGY AND SYSTEMATICS OF *CRYPTOSPORIDIUM PARVUM*

Una M. Morgan,[a]* Lihua Xiao,[b] Ronald Fayer,[c] Altaf A. Lal,[b] and R.C. Andrew Thompson [a]

aWorld Health Organization Collaborating Centre for the Molecular Epidemiology of Parasitic Infections and State Agricultural Biotechnology Centre, Division of Veterinary and Biomedical Sciences, Murdoch University, Murdoch, WA, 6150, Australia; bDivision of Parasitic Diseases, Centers for Disease Control and Prevention, Atlanta, Georgia, 30341, USA; cImmunology and Disease Resistance Laboratory, Agricultural Research Service, USDA, Beltsville, Maryland, 20705 USA;

1 INTRODUCTION

At present, at least 10 species of *Cryptosporidium* are regarded as valid on the basis of differences in oocyst morphology, site of infection, and vertebrate class specificity: *C. muris* which infects rodents, *C. andersoni* which infects cattle; *C. parvum* which infects humans and other mammals; *C. meleagridis* and *C. baileyi* in birds; *C. serpentis* in reptiles; *C. saurophilum* in lizards; *C. nasorum* in fish, *C. wrairi* from guinea pigs and *C. felis* in cats[1,2]. Molecular characterisation studies have detected considerable evidence of genetic heterogeneity between isolates of *Cryptosporidium* from different species of vertebrate and within *C. parvum*, and there is now mounting evidence suggesting that a series of host-adapted cryptic species of the parasite exist[1,2].

1.1 Molecular Epidemiology of Cryptosporidium parvum

Antigenic differences[3-5], as well as differences in virulence and pathogenesis, infectivity, and drug sensitivity have been identified within isolates of *C. parvum*[6-8].

1.1.1 Cryptosporidium in Humans and Domestic Livestock - 'Human' and 'Cattle' Genotypes. There is now strong evidence from genetic and biological characterisation studies that there are two distinct genotypes of *Cryptosporidium* infecting humans: a 'human' genotype which to date has been found only in humans and a 'cattle' genotype which is found in domestic livestock such as cattle, sheep, goats etc as well as humans indicating zoonotic transmission[9-40].

Genetic diversity in *C. parvum* has been identified using isoenzyme analysis which clearly differentiated between human and animal isolates, with some human isolates exhibiting animal profiles as a result of zoonotic transmission[10,11,41]. The difference between human and animal *C. parvum* isolates was subsequently confirmed by RAPD (Random Amplified Polymorphic DNA) analysis[13,15,42,43]. However, due to the low numbers of oocysts frequently recovered from environmental and faecal samples, and the presence of contaminants, the majority of genetic characterisation studies have utilised parasite-specific PCR primers to overcome these problems.

Sequence and/or PCR-RFLP (Restriction Fragment Length Polymorphism) analysis of the 18S rDNA gene[13,16,18,23,35-38] and the more variable internal transcribed rDNA spacers (ITS1 and ITS2)[13,19], the acetyl-Co A synthetase gene[18], the COWP gene[26,27,38], the dihydrofolatereductase-thymidylate synthase (*dhfr*) gene[14,23,33], the 70 kDa HSP70[32], the thrombospondin-related adhesion protein (TRAP-C1 and TRAP-C2) genes[24,29,30,38] and an unidentified genomic fragment[12] have all confirmed the genetic distinctness of the 'human' and 'cattle' genotypes.

A recent multilocus approach analysed 28 isolates of *Cryptosporidium* originating from Europe, North and South America and Australia[27]. PCR-RFLP analysis of the polythreonine [poly(T)] and COWP gene, TRAP-C1 gene and ribonucleotide reductase gene (RNR), and genotype specific PCR analysis of the rDNA ITS 1 region, clustered all the isolates into two groups, one comprising both human and animal isolates and the other comprising isolates only of human origin[27]. PCR-RFLP analysis of the poly(T) and COWP gene, RNR and PCR analysis of the 18S rDNA gene was also conducted on *C. parvum* isolates from AIDS patients[34]. Five of the patients tested exhibited the 'human' genotype and 2 exhibited the 'cattle' genotype. In both studies, neither recombinant genotypes nor mixed infections were detected[34]. Another study analysed 211 faecal samples "positive" for *Cryptosporidium* by microscopy using PCR-RFLP analysis of 18S rRNA, COWP, and TRAP-C1 gene fragments[38]. Of the samples analysed 38% were the human genotype and 62% were the cattle genotype[38]. The human genotype was detected in a significantly greater proportion of the samples with larger numbers of oocysts and the cattle genotype was detected in a significantly greater proportion of the samples with small numbers of oocysts, indicating that there may be differences in virulence between the two genotypes. However, this needs to be examined further. In this study, there were no significant differences in the distribution of the genotypes by patient sex and age. The distribution of the genotypes however, was significantly different both in patients with a history of foreign travel and in those from different regions in England[38].

In foodborne, waterborne and day care outbreaks of cryptosporidiosis, oocysts of both the human and bovine genotypes of *Cryptosporidium* have been identified, with the human genotype identified more frequently[24, 30, 36, 38,39, 44]. Outbreaks caused by the bovine genotype have all been epidemiologically linked to contamination from or direct contact with animals, such as the Maine apple cider outbreak in 1995, the British Colombia outbreak in 1996, the Pennsylvania rural family outbreak in 1997 and the Minnesota zoo outbreak in 1997[30]. Results of these studies were also very useful in clarifying the source of contamination in outbreaks, such as the massive one in Milwaukee in 1993, which was probably caused by *Cryptosporidium* of human origin contaminating the water supply[24,30].

Studies to date have indicated substantial genetic differences between the 'human' and 'cattle' genotypes but little variation within these genotypes. Minor differences within the human genotype isolates have been identified in the 18S rRNA[35], TRAP-C2[24,30] and poly(T) genes[34]. High resolution typing tools are needed to identify outbreaks and to track infection and contamination sources. However, preliminary analysis of currently available *Cryptosporidium* databases has indicated that most microsatellite sequences detected were AT-rich microsatellites of low complexity[40]. Recently microsatellite analysis on 94 *C. parvum* human and animal isolates, was used to differentiate the 'human' genotype into 2 subgenotypes and the cattle genotype into 4 subgenotypes[40]. Some subgenotypes showed a wide geographical distribution, whereas others were restricted to specific regions. Another study characterised 9 microsatellite loci and identified 2 subgenotypes within the human genotype and 2 subgenotypes within the cattle genotype[46]. A number of subgenotypes have also been identified within the human and cattle genotypes using sequence analysis of the hsp70 locus[Xiao et al. unpublished observations] Additional loci need to be characterised in order to obtain greater intragenotype variation. Recent sequence analysis of a highly polymorphic *C. parvum* gene encoding a 60-kilodalton glycoprotein and characterization of its 15-and 45-kilodalton zoite surface antigen products will prove useful in the future for haplotyping and fingerprinting isolates and for establishing meaningful relationships between *C. parvum* genotypes and phenotypes[47].

1.1.2   Additional C. parvum-like Genotypes/Cryptic Species. A number of additional genetically distinct genotypes/cryptic species have been identified. Recent research, genetically characterising isolates of *C. parvum* from mice (*Mus musculus*) in Australia, the UK, and Spain using sequence analysis of the 18S rRNA, ITS, *dhfr* and AcetylCo A loci as well as RAPD analysis has revealed that these isolates carry a distinct genotype referred to as the 'mouse' genotype[18,19,22,36]. Interestingly, some of the mice were also infected with the 'cattle' genotype indicating that they might serve as reservoirs of infection for humans and other animals. Oocysts of the 'mouse' genotype were also identified from a large footed mouse-eared bat *(Myotus adversus)*, extending the host range of this genotype[22]. Pigs have also been shown to be infected with a genetically distinct host- adapted form of *Cryptosporidium*[18,19,21,36,48]. Little is known about the prevalence of *Cryptosporidium* in marsupials. *Cryptosporidium* infections have been reported in southern brown bandicoots (*Isoodon obesulus*), a hand-reared juvenile red kangaroo *(Macropus rufus)* from South Australia and a Tasmanian wallaby (*Thylogale billardierii*)[49]. Genetic analysis of marsupial isolates at the 18S rDNA, ITS, dhfr and hsp70 loci have all confirmed their genetic identity, and distinctness from other all other genotypes of *C. parvum*[23,36].

Genetic analysis of *C. parvum*-like isolates from dog (*Canis familiaris*) isolates from the USA and Australia and from ferret (*Mustela furo*) isolates at the 18S rDNA and HSP70 loci have also revealed distinct genotypes[36,50]. Recently, a monkey genotype has also been identified based on the analysis of the 18S rRNA, HSP70 and COWP genes[36]. As expected, this genotype is most related to the human genotype. As more isolates of *Cryptosporidium* from other animal species are analysed genetically, it is likely that new additional genotypes will be identified.

1.1.3   Infectivity of Cryptosporidium Genotypes for Immunocompromised Hosts.
Few genotyping studies have been conducted on isolates of *Cryptosporidium* from immunocompromised patients[12, 30,34,51,52]. A recent study genotyped 10 *Cryptosporidium* isolates from HIV-infected individuals at the 18S rDNA locus[51]. In this study, 1 isolate exhibited the 'cattle' genotype, 5 isolates exhibited the 'human' genotype, 3 were infected with *C. felis* and 1 exhibited the newly identified 'dog' genotype[51]. For some patients, multiple specimens collected over 12 months were available and in these cases the same *Cryptosporidium* genotype persisted throughout the course of the patient's infection[51]. In another study, *Cryptosporidium* isolates from HIV-infected individuals from Switzerland, Kenya and the USA were analysed at 3 genetic loci: the 18S rDNA, HSP-70 and AcetylCoA synthethase genes[52]. The results revealed that the majority of the patients (64%) were infected with the 'human' and 'cattle' *C. parvum* genotypes. However, a number of patients were infected with *C. felis* (27%) and *C. meleagridis* (9%)[52]. These results indicate that immunocompromised individuals are susceptible to a wide range of *Cryptosporidium* species and genotypes and clearly host-factors must play a role in controlling susceptibility to these divergent parasites. Future studies on a larger number of AIDS patients with more extensive clinical information is required in order to understand the full public health significance of *Cryptosporidium* species and genotypes

in immunocompromised hosts.

1.2 How do we Define a Species?

Achieving a sound taxonomy for the genus *Cryptosporidium* is essential to understanding the epidemiology and transmission of cryptosporidial infections, and for controlling outbreaks of the disease. The picture that is emerging as a result of molecular studies clearly indicates that the species-level taxonomy of the genus does not reflect molecular phylogenetic analyses nor epidemiological data, and warrants reappraisal. The classical definition of species as groups of interbreeding natural populations reproductively isolated from other groups is difficult to apply to protozoa. As a result, species within the Apicomplexa have frequently been described based on combinations of morphological features detected by light microscopy or EM, unique life cycles and host specificity. However, difficulties in species identification arise when the size, shape or internal structures of oocysts of one species cannot be distinguished from those of another species. Such is the case with the relatively small oocysts associated with species of *Cryptosporidium*. When distinct morphological differences are lacking, the use of these criteria varies with individual parasites, and the determination of what is a species is subject to the interpretations of researchers in a particular field[1,2].

1.3 Evidence for Multiple Species within *C. parvum*

There are distinct biological and genetic differences between isolates of *C. parvum* from humans, cattle, pigs, dogs, mice ferrets and marsupials[1,2,23,32,36]. Most of the genotypes identified to date with the exception of the 'cattle' genotype appear to be host-specific[1,2]. In addition, the 'human', 'pig' and 'marsupial' genotypes do not appear to readily infect mice and the 'human' genotype does not appear to be infectious for cattle[1,2,24,34]. Phylogenetic analysis of 18S rDNA data indicates that *Cryptosporidium* can be divided into 2 broad groups: the gastric parasites (*C. muris, C. andersoni* and *C. serpentis*) and the intestinal parasites (*C. parvum, C. wrairi, C. felis, C. meleagridis, C. saurophilum*). Phylogenetic studies at several different loci have shown that *C. parvum* is not monophyletic as *C. meleagridis* and *C. wrairi* cluster within the "*C. parvum*" group. For example, a recent study has shown that the genetic similarity between the 'dog' genotype and the 'human/'cattle' genotypes was 97%[50]. This is on par with the genetic similarity between 2 accepted species of *Cryptosporidium*, namely *C. muris* and *C. serpentis* (97%) and less than the similarity between *C. parvum* ('cattle'/'human'), *C. wrairi* and *C. meleagridis* (99-98.4%). Similarly the genetic similarity between the *C. parvum* 'cattle'/'human' genotypes and the 'pig' and 'marsupial genotypes at the 18S rRNA locus was 98.7% and 98.5% respectively. This is less than the observed genetic similarity of 99.5%-99.3% between *C. parvum* ('human'/'cattle') and *C. wrairi*[23]. Biological and phylogenetic data coupled with the apparent lack of recombination between genotypes even though mixed genotype infections are known to occur, all provide a strong case that many of these '*C. parvum*-like' genotypes are in fact cryptic species[1,2]. Clarification of the taxonomy of *Cryptosporidium* will provide more precise information in assessing the public health significance of certain *Cryptosporidium* species and genotypes, and therefore promote the identification of risk factors and the development of preventative measures[1,2].

References

1. U. M. Morgan, L. Xiao, R. Fayer, A. A. Lal and R. C. A. Thompson, *Int. J. Parasitol.*, 1999, **29**, 1733.
2. L. Xiao, U. M. Morgan, R. Fayer, R. C. A. Thompson and A. A. Lal, *Parasitol. Today*, 2000, **16**: 287.
3. V. McDonald, R. M. A. Deer, J. M. S. Nina, S. Wright, P. L. Chiodini and K. P. W. J. McAdam, *Parasite Immunol.*, 1991, **13**, 251.
4. G. L. Nichols, J. McLauchlin and D. Samuel, *J. Protozool.*, 1991, **38**, 237S.
5. K. Griffin, E. Matthai, M. Hommel, J. C. Weitz, D. Baxby, C and A. Hart, *J. Protozool. Res.*, 1992, **2**, 97.
6. J. R. Mead, R. C. Humphreys and D. W. Sammons, *Infect. Immun.*, 1990, **58**, 2071.
7. W. L. Current and N. C. Reese, *J. Protozool.*, 1986, **33**, 98.
8. R. Fayer and B. L. P. Ungar, *Microbiol. Rev.*, 1986, **50**, 458.
9. Y. R. Ortega, R. R. Sheehy, V. A. Cama and K. K. Oishi, *J. Protozool*, 1991, **38**, 40S.
10. F. M. Awad-El-Kariem, H. A. Robinson, D. A. Dyson, D. Evans, S. Wright, M. T Fox and V. M. McDonald, *Parasitol.*, 1995, **110**, 129.
11. F. M. Awad-El-Kariem, H. A. Robinson, D. A. Dyson, D. Evans, S. Wright, M. T. Fox, F. Petry, V. McDonald, D. Evans and D. Casemore, *Parasitol. Res.*, 1998, **84**, 297.
12. A. M. Bonnin, N. Fourmaux, J. F. Dubremetz, R. G. Nelson, P. Gobet, G. Harly, M. Buisson, D. Puygauthier-Toubas, F. Gabriel-Pospisil, M. Naciri and P. Camerlynck, *FEMS Microbiol. Lett.*, 1996, **137**, 207.
13. M. Carraway, S. Tzipori and G. Widmer, *Appl. Environ. Microbiol.*, 1996, **62**, 712.
14. C. L. Gibbons, B. G. Gazzard, M. A. A. Ibraham, S. Morris-Jones, C. S. L. Ong and F. M. Awad-El-Kariem. *Parasitol. Int.*, 1998, **47**, 139.
15. U. M. Morgan, C. C. Constantine, P. O'Donoghue, B. P. Meloni, P. A. O'Brien and R. C. A. Thompson, *Am. J. Trop. Med. Hyg.*, 1995, **52**, 559.
16. U. M. Morgan, C. C. Constantine, D. A. Forbes and R. C. A. Thompson. *J. Parasitol.*, 1997, **83**, 825.
17. U. M. Morgan, C. C. Constantine and R. C. A. Thompson, *Parasitol. Today.* 1997, **13**, 488.
18. U. M. Morgan, K. D. Sargent, P. Deplazes, D. A. Forbes, F. Spano, H. Hertzberg, A. Elliot and R. C. A. Thompson, *Parasitol.*, 1998, **117**, 31.
19. U. M. Morgan, K. D. Sargent, P. Deplazes, D. A. Forbes, F. Spano, H. Hertzberg, A. Elliot and R. C. A. Thompson, *Parasitol.*, 1999, **118**, 49.
20. U. M. Morgan, L. Pallant, B. W. Dwyer, D. A. Forbes, G. Rich and R. C. A. Thompson, *J. Clin. Microbiol.*, 1998, **36**, 995.
21. U. M. Morgan, R. Buddle, A. Armson and R. C. A. Thompson, *Aust. Vet. J.*, 1999, **77**, 44.
22. U. M. Morgan, A. P. Sturdee, G. Singleton, M. S. Gomez, M. Gracenea, J. Torres, S. G. Hamilton, D. P. Woodside and R. C. A. Thompson, *J. Clin. Microbiol.*, **37**, 1302.
23. U. M. Morgan, P. T. Monis, R. Fayer, P. Deplazes and R. C. A. Thompson, *J. Parasitol.*, 1999, **85**, 1126.

24. M. M. Peng, L. Xiao, A. R. Freeman, M. J Arrowood, A. A. Escalante, A. C. Weltman, C. S. Ong, W. R. MacKenzie, A. A. Lal and C. B. Beard, *Emerg. Infect. Dis.*, 1997, **3**, 567.
25. P. A. Rochelle, E. M. Jutras, E. R. Atwill, R. De Leon and M H. Stewart, *J. Parasitol.*, 1999, **85**, 986.
26. F. Spano, L. Putignani, J. McLauchlin, D. P. Casemore and A. Crisanti, *FEMS Microbiol. Let.*, 1997, **150**, 209.
27. F. Spano, L. Putignani, A. Crisanti, P. Sallicandro, U. M. Morgan, S. M. Le Blancq, L. Tchack, S. Tzipori and G. Widmer, *J. Clin. Microbiol.*, 1998, **36**,3255.
28. F. Spano, L. Putignani, S. Naitza, C. Puri, S. Wright and A. Crisanti, *Mol. Biochem. Parasitol.*, 1998, **92**, 147.
29. F. Spano, L. Putignani, S. Guida and A. Crisanti, *Exp. Parasitol.*, 1998, **90**, 195.
30. I. M. Sulaiman, L. Xiao, C. Yang, L. Escalante, A. Moore, C. B. Beard, M. J. Arrowood and A. A. Lal, *Emer. Infect. Dis.*, 1998, **4**, 681.
31. I. M. Sulaiman, A. A. Lal, M. J. Arrowood and L. Xiao, *J. Parasitol.*, 1999, **85**, 154.
32. I. M. Sulaiman, U. M. Morgan, R. C. A. Thompson, A. A. Lal and L. Xiao, *App. Environ. Microbiol.*, 2000, **66**, 2385.
33. J. R. Vasquez, L. Gooze, K. Kim, J. Gut, C. Petersen and R. G. Nelson, *Mol. Biochem. Parasitol.*, 1996, **79**, 153.
34. G. Widmer, S. Tzipori, C. J. Fichtenbaum and J. K. Griffiths, *J. Infect. Dis.*, 1998, **178**, 834.
35. L. Xiao, I. Sulaiman, R. Fayer and A. A. Lal, *Memorias do Instituto Oswaldo Cruz*, 1998, **93**, 687.
36. L. Xiao, U. Morgan, J. Limor, A. Escalante, M. Arrowood, W. Schulaw, R. C. A. Thompson, R. Fayer and A. A. Lal, *Appl. Environ. Microbiol.*, 1999, **65**, 3386.
37. L. Xiao, L. Escalante, C. F. Yang, I. Sulaiman, A. A. Escalante, R. J Montali, R. Fayer and A. A. Lal, *Appl. Environ. Microbiol.*, 1999, **65**,1578.
38. J. McLauchlin, S. Pedraza-Diaz, C. Amar-Hoetzeneder and G. L. Nichols, *J. Clin. Microbiol.* 1999, **37**, 3153.
39. C. S. Ong, D. L. Eisler, S. H. Goh, J. Tomblin, F. M. Awad-El-Kariem, C. B. Beard, L. Xiao, I. Sulaiman, A. Lal, M. Fyfe, A. King, W. R. Bowie and J. L. Isaac-Renton, *Am. J. Trop. Med. Hyg.*, 1999, **61**, 63.
40. S. Caccio, W. Homan, R. Camilli, G. Traldi, T. Kortbeek and E. Pozio,. *Parasitol.*,.2000, **120**, 237.
41. B. W. Ogunkolade, H. A. Robinson, V. McDonald, K. Webster and D. A. Evans, *Parasitol. Res.*, 1993, **79**, 385.
42. M. Q. Deng and D. O. Cliver, *Appl. Environ. Microbiol.*, 1998, **64**, 1954.
43. K. V. Shianna, R. Rytter and J G. Spanier, *Appl. Environ. Microbiol.*, 1998, **64**, 2262.
44. M. Q. Deng and D. O. Cliver, *Int. J. Food Microbiol.*, 2000, **54**, 155.
45. S. Patel, S. Pedraza-Diaz, J. McLauchlin and D. P. Casemore, *Commun. Dis. Pub. Hlth.*, 1998, **1**, 232.
46. A. E. Aiello, L. H. Xiao, J. R Limor, C. Liu, M. S. Abrahamsen and A. A. Lal, *J. Eukar. Microbiol.*, 1999, **46**, 46S.
47. W. B. Strong, J. Gut, R. G. Nelson, *Inf. Immun.*, 2000, **68**, 4117.
48. M. Pereira, E. R. Atwill, M. R. Crawford and R. B Lefebvre, *Appl. Environ. Microbiol.* 1998, **64**, 1584.

49. P. J. O'Donoghue, *Int. J. Parasitol.,*1995, **25**, 139.
50. U. M. Morgan, L. Xiao, P. Monis, A. Fall, P. J. Irwin, R. Fayer, K. Denholm, J. Limor, A. Lal and R. C. A. Thompson, *App. Environ. Microbiol.,* 2000a, **66**, 2220.
51. N. J. Pieniazek, F. J Bornay-Llinares, S. B. Slemenda, A. J. da Silva, I. N. Moura, M. J. Arrowood, O. Ditrich and D. G. Addiss, *Emerg. Infect. Dis.,* 1999, **5,** 444.
52. U. M. Morgan, R. Weber, L. Xiao, I. Sulaiman, R. C. A. Thompson, W. Ndiritu, A. A. Lal, A. Moore and P. Deplazes, *J. Clin. Microbiol.,* 2000, **38**, 1180.

MOLECULAR AND PHENOTYPIC ANALYSIS OF *CRYPTOSPORIDIUM PARVUM* OOCYSTS OF HUMAN AND ANIMAL ORIGIN

G. Widmer, X. Feng, D. Akiyoshi, S.M. Rich, B. Stein, S. Tzipori

Division of Infectious Diseases
Tufts University School of Veterinary Medicine
North Grafton, Massachusetts 01536

1 INTRODUCTION

Genotypic and phenotypic characterization of isolates of *Cryptosporidium parvum* have identified two groups, designated genotype 1 and 2[1-3]. Oocysts belonging to different genotypes vary not only with respect to their genetic profile, but also display significant phenotypic differences. Although both genotypes cause cryptosporidosis in humans, the infectivity of genotype 1 and 2 oocysts for laboratory animals varies [2,4,5]. Because of the difficulty in propagating genotype 1 *C. parvum*, this genotype has been poorly studied. It is unknown to what extent the observed properties of certain genotype 1 isolates apply to the entire group. It is also unclear whether genotype 1 and 2 oocysts differ in their sensitivity to water disinfectants.

2 *CRYPTOSPORIDIUM PARVUM* IS A HETEROGENEOUS SPECIES

Using primarily genetic methods, several laboratories have identified two main subgroups within the species *C. parvum*[6]. Restriction fragment length polymorphism (RFLP) analysis has been useful for this work as PCR amplification of polymorphic regions can be performed with DNA extracted directly from stool samples, from water pellets or from a small number of oocysts[7,8]. Polymorphic restriction sites were identified within known gene sequences, primarily located in protein coding regions [5, 9-11]. Because of the nature of the RFLP technique, and the fact that coding regions were targeted, most of these studies did not reveal any additional heterogeneity beyond the two main genotypes.

Nucleotide sequence analysis at selected loci revealed an additional level of heterogeneity within each genotype. Multiple RFLP profiles were identified within a fragment of the β-tubulin gene spanning an intron and confirmed by sequence analysis[12]. Additional genotypes, some of which are associated with particular host species, were also described[13,14].

A survey of multi-locus genotypes based on markers located on different chromosomes[15] failed to identify recombinant genotypes. Only two RFLP profiles were obtained from different hosts and geographic origins. This observation suggests that *C. parvum* isolates designated as genotypes 1 and 2 do not recombine and should perhaps be considered separate species. Alternative explanations include spacial separation within the host and/or a selective disadvantage of recombinant genotypes.

3 ANIMAL PROPAGATION OF *C. PARVUM*

The study of phenotypic properties of *C. parvum* is hampered by the lack of in vitro propagation and cryopreservation methods. These limitations make it difficult to generate genetically defined parasite lines and maintain them over extended periods of time to allow the study of different properties. In the absence of such tools, *C. parvum* isolates are propagated in laboratory animals, usually calves[16]. It is unknown to what extent transmission of isolates between different hosts affects the parasite population. This concern raises the possibility that the composition of the original isolate may change following passage to a different host.

Maintenance of *C. parvum* genotype 2 in animals has been practiced for many years. This genotype readily infects calves and laboratory rodents. In contrast, genotype 1 does not normally infect calves nor neonatal mice[2,4,5]. Some genotype 1 isolates also fail to infect immunosuppressed mice (Deary, Akiyoshi and Tzipori, unpublished). Therefore, we explored the feasibility of propagating *C. parvum* genotype 1 in piglets. Gnotobiotic piglets were infected with four genotype 1 isolates, three of human origin and one isolated from a captive macaque. Three initial attempts at propagating genotype 1 resulted in a conversion of genotype 1 to genotype 2 in the course of successive pig passages (Fig. 1). It is unknown whether this change was caused by a contamination of the isolates with oocysts of genotype 2, or whether the original oocyst samples already harbored a genotype 2 subpopulation. Whichever the origin of type 2 oocysts is, this observation suggests the existence of a selective mechanism by which growth of these parasites is favored in the pig. In a subsequent attempt at propagating a human type 1 isolate, stringent conditions to prevent possible cross-contamination of oocyst stocks were implemented. In this experiment, in a series of 8 pig infections over a 10 month period, the original type 1 RFLP profile of the isolate, termed NEMC1, was maintained. Although expensive and labor-intensive, propagation of isolate NEMC1 has led to a better characterization of a type 1 isolate than had been possible previously.

4 PHENOTYPIC PROPERTIES OF GENOTYPE 1 AND 2 OOCYSTS

In contrast to genotype 2 isolates propagated in calves, isolate NEMC1 was not infectious to immunodeficient mice (C57BL/6 interferon-γ knock-out). This property was unexpected and additional genotype 1 isolates will be tested in this model[17] in order to assess their infectivity.

Oocysts from several genotype 1 and 2 isolates were examined for their stability at room temperature. Purified oocysts were suspended in sterile water and the number of intact oocysts counted over a period of three weeks. Interestingly, the isolates segregated into two groups (Fig. 2). One group comprising several genotype 2 isolates

**9897
human**

**PC2
macaque**

**PC1HIV1
human**

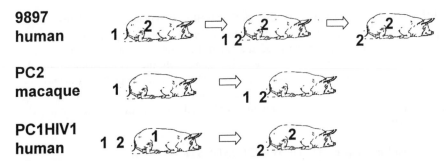

Figure 1 *Propagation of three genotype 1 isolates in gnotobiotic piglets.* The isolate designation and host of origin of the isolates is shown left. Piglets were orally infected with oocysts. Oocysts excreted in the feces were purified and genotyped using RFLP markers[15]. The genotype identified is shown behind each animal. In addition to fecal collection, oocysts were obtained from some animals' intestines after euthanasia. The genotype of these samples is indicated on the animal. Isolate 9897 (2nd passage), PC2(2nd passage) and PC1HIV1 (1st passage) excreted oocysts of mixed genotype 1 and 2 profile as indicated. Human isolate PC1HIV1 and PC2 were a gift from Dr. Cynthia Chappell, University of Texas, Houston, and Dr. Keith Mansfield, New England Regional Primate Center, Southboro, Massachusetts, respectively.

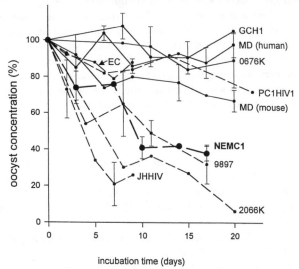

Figure 2 *Oocyst decay at room temperature.* Oocysts from nine *C. parvum* isolates were surface-sterilized and incubated in sterile water. Replicate counts of intact oocysts were determined over a period of up to 22 days. Mean counts, expressed as a percentage of the count at day 0, and selected standard deviations are shown. Dashed lines, genotype 1; continuous lines, genotype 2. Isolates JHHIV, EC and MD (human) were provided by Dr. Cynthia Chappell. Oocysts from the Moredun (MD) isolate propagated in mice and in human volunteers are included. Isolates 0676K and 2066K are from HIV positive individuals[5]. Isolate 9897 is from an HIV-negative individual, GCH1 is from a calf and NEMC1 from a pig.

of different host origin (human, mouse, calf) as well as two genotype 1 isolates displayed a stability typically associated with *C. parvum* oocysts[18]. In contrast the remaining genotype 1 oocysts decayed at a much faster rate, with a half-life of approximately 7 days. Oocysts of isolate NEMC1 fell into the fast decaying group. At 4°C, the stability of NEMC1 oocysts was similar to that of genotype 2 oocysts; no significant drop in oocyst counts were observed over a period of five weeks. Scanning electron microscopy did not reveal any morphological differences between genotype 1 and 2 oocysts.

5 RELEVANCE TO THE WATER INDUSTRY

It is conceivable that genetic factors play a role in determining the resistance of oocysts to water disinfectants. The difficulty in generating significant numbers of genotype 1 oocysts has hampered comparative disinfection studies with both genotypes. Properties of genotype 1 oocysts relevant to the water industry remain to be elucidated. The exclusive reliance on genotype 2 oocysts for disinfection studies may not reveal the full extent of phenotypic variability within *C. parvum*. Additional complications are likely to arise from the fact that there may be incomplete correspondence between parasite genotype and its phenotype. For instance, in the oocyst stability data shown in Fig. 2, the genotype was not entirely predictive of stability, as both genotypes were found among the high stability group.

In summary, technical obstacles have delayed research on many important aspects of *Cryptosporidium*. Of practical interest to environmental and clinical applications is the identification of markers associated with specific phenotypic properties. Culture methods capable of maintaining *C. parvum* in vitro will open many new research possibilities, including the search for genetic markers determining virulence, environmental resistance and host specificity. In the meantime, propagation in laboratory animals is the only alternative for generating oocysts.

References

1. M Carraway, S Tzipori and G Widmer, *Appl. Environ. Microbiol.* 1996, **62**, 712.
2. MM Peng, , L Xiao, AR Freeman, MJ Arrowood, AA Escalante, AC Weltman, CSL Ong, WR MacKenzie, AA Lal, CB Beard,. *Emerg Infect Dis* 1997,**3**,567.
3. BW Ogunkolade, HA Robinson, V McDonald, K Webster, DA Evans, *Parasitol. Res.* 1993, **79**, 385.
4. FM Awad-El-Kariem, HA Robinson, F Petry, V McDonald, D Evans, and D Casemore, *Parasitol. Res.* 1998, **84**, 297.
5. G Widmer, S Tzipori, CJ Fichtenbaum and JK Griffiths, *J. Infect. Dis.* 1998, **178**, 834.
6. G Widmer. *Adv. Parasitol.* 1998, **40**, 224.
7. PA Rochelle, R De Leon, MH Stewart, LR Wolfe, *Appl. Environ. Microbiol.* 1997, **63**, 106.
8. SD Sluter, S Tzipori, G Widmer, *Appl. Microbiol. Biotechnol.*, 1997, **48**, 325.
9. M Carraway, S Tzipori and G Widmer *Infect. Immun.* 1997, **65**, 3958.
10. F Spano, L Putignani, J McLaughlin, DP Casemore, A Crisanti, *FEMS Microbiol Letters* 1997, **152**, 209.

11. F Spano, L Putignani, S Guida, A Crisanti, *Exp. Parasitol.* 1998, **90**, 195.
12. G Widmer, L Tchack, CL Chappell, S Tzipori, *Appl. Environ. Microbiol.* 1998, **64**, 4477.
13. UM Morgan, P Deplazes, DA Forbes, F Spano, H Hertzberg, KD Sargent, A Elliot, RCA Thompson, *Parasitol.* 1999, **118**, 49.
14. NJ Pieniazek, FJ Bornay-Llinares, SB Slemenda, AJ da Silva, INS Moura, MJ Arrowood, O Ditrich, DG Addiss, *Emerg. Infect. Dis.* 1999, **5**, 444.
15. F Spano, L Putignani, A Crisanti, A Sallicandro, UM Morgan, SM Le Blancq, L Tchack, S Tzipori, G Widmer, *J. Clin. Microbiol.* 1998, **36**, 3255.
16. S Tzipori, *Adv. Parasitol.* 1998, **40**, 188.
17. C Theodos, K Sullivan, T Gull, S Tzipori, *Infect. Immun.* 1997, **65**, 4761.
18. S Yang, MC Healey, C Du, *FEMS Immunol Med Microbiol* 1996, **13**, 141.

COMPLYING WITH THE NEW *CRYPTOSPORIDIUM* REGULATIONS

P. Jiggins

Drinking Water Inspectorate
Ashdown House
123 Victoria Street
LONDON SW1E 6DE

1 INTRODUCTION

For some time it has been recognised that *Cryptosporidium* in water supplies represent a significant risk to health. In the last twelve years there have been at least 25 outbreaks associated with mains water supplies in England and Wales (1). Almost all of these outbreaks have been associated with failures or inadequacies in the water treatment process or its operation. A great deal of progress to minimise the risk to public health has been made in the years following the first major water-borne outbreak affecting parts of Swindon in 1981. Three reports have been published by the Expert Groups chaired initially by the late Sir John Badenoch (2) (3) and more recently by Professor Ian Bouchier which contain many recommendations to improve aspects of water treatment.

Although progress had been made to develop appropriate sampling and analytical techniques for the detection of *Cryptosporidium*, recovery rates from water samples have remained poor. Routine monitoring of water supplies was not considered feasible. However in recent years there have been a number of significant technical developments in sampling and analytical techniques that have resulted in much improved and consistent levels of recovery.

Within the present Water Supply (Water Quality) Regulations 1989 (as amended) there is no specific numerical standard for *Cryptosporidium*, however there is a general requirement that drinking water "does not contain any element, organism or substance....at a concentration or value which would be detrimental to public health. Water that contained sufficient *Cryptosporidium* oocysts to cause illness would therefore be unwholesome. Further, water that is likely to cause, or has caused cryptosporidiosis is unfit for human consumption. The supply of such water would be an offence under Section 70 of the Water Industry Act 1991. The Drinking Water Inspectorate has taken forward one prosecution under this power in relation to an incident believed to have arisen from a water-borne infection by *Cryptosporidium*. The water company was acquitted because the epidemiological study linking the outbreak to the water supply was ruled inadmissible as evidence.

Against this background Ministers considered that further regulations where needed to provide additional protection to public health by reducing the risk of *Cryptosporidium* being found in drinking water supplies. The new regulations came into force on 30 June 1999 after a period of consultation.

2 THE WATER SUPPLY (WATER QUALITY) (AMENDMENT) REGULATIONS 1999 (4)

The new regulations require water companies to carry out a risk assessment for each of their treatment works to establish whether there is a significant risk that water would be supplied that contained *Cryptosporidium* oocysts above the new standard. A guidance note has been issued by the Secretary of State which describes the factors which should be assessed and the format of the risk report (5).

For the purposes of the new regulations certain types of site should in all cases be classified as constituting a significant risk:

a) direct abstractions from rivers or streams (where average storage is seven days or less);

b) groundwaters with rapid surface water connections as indicated by the presence of faecal coliform bacteria;

c) supplies associated with previous outbreaks of cryptosporidiosis, where no changes to treatment have been made.

Water companies have completed their initial risk assessments and identified a total of 308 sites where there is a significant risk of exceeding the standard. Approximately 135 of these sites relate to direct abstractions and 173 sites relate to vulnerable groundwaters.

At sites where a significant risk is identified water companies must use a treatment process to secure that the water supplied meets the standard of less than one *Cryptosporidium* oocyst per 10 litres as an average value. At such sites the water company must verify that the process meets the standard by continuos monitoring, unless the process incorporates approved treatment capable of continuously removing or retaining particles greater than one micron.

The Secretary of State has also published four Standard Operating Protocols to specify the requirements for sampling (6), analysis (7), validation of new methods (8) and an inter-laboratory proficiency scheme (9).

The regulations establish a new criminal offence of allowing levels of *Cryptosporidium* oocysts above the standard in treated water leaving a treatment works or not complying with the sampling or analytical requirements.

3 SAMPLING AND TRANSPORTATION REQUIREMENTS

The conditions of sampling and examination have been specified to permit the use of analytical results as evidence in a Court of Law. The Standard Operating Protocol provides a detailed specification for sampling equipment and the associated materials required for the secure transportation of samples to an approved laboratory. The sampling cabinet and associated security items must be of the type specified within the protocol and obtained from the Drinking Water Inspectorate (DWI) approved suppliers. The protocol specifies the requirements for the:

a) sampling cabinet

b) security measures to be taken during transportation of samples;

c) security tags, padlocks and evidence bags used;

d) documentation of sampling data.

The sampling cabinet houses the sampling collection device and associated pumping, valving and monitoring equipment. The sampling collection device can be quickly removed by the use of swagelock fittings on flexible hose at the inlet and outlet, the remainder of the sample line is fixed. Sampling cabinets are currently available from two approved suppliers, an example of an approved cabinet is illustrated in Figure 1.

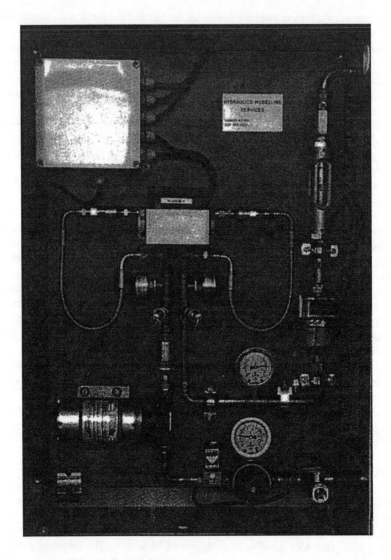

Figure 1 *Sampling cabinet (Prototype 6)*

Once the sampling equipment has been installed at the site it will be subject to a DWI audit, prior to approval for use. The audit will check that the installation conforms fully to the specification detailed in the protocol. During the audit the security of the installation will be verified by the addition of security tags, paints and tapes over joints and connections.

The protocol also specifies the procedure for collection of the sampling device on completion of the sampling period and the installation of the replacement filter. The sample must be kept in secure conditions during transit to the laboratory with appropriate documentation to demonstrate continuity during any transfer points. The new regulations require continuous sampling to be undertaken at a flow rate of at least 40 litres per hour on average in each sampling period, with allowance of one hour given for changing the sampling device. For sites which supply water on a continuous basis the sampling collection device must be changed at least once a day, which means in practice that samples volumes should exceed 920 litres. Where sites supply water on an intermittent basis the sampling device need not be changed daily but on the day that at least 200 litres has been sampled.

4 ANALYTICAL REQUIREMENTS

During the consultation period for the new regulations DWI let two contracts to evaluate the feasibility of continuous monitoring for *Cryptosporidium* in treated water supplies. The specific objectives of the contracts were to evaluate whether a filter device would allow monitoring of 1000 litres of treated water between changes. The device would have to be sufficiently robust to withstand continuous operation and the required operating pressures. It should allow subsequent recovery of *Cryptosporidium* oocysts at a satisfactory level. The contract also sought to demonstrate whether there was a laboratory analytical method which was robust, repeatable and capable of achieving an acceptable recovery. The study (10) concluded that one sampling device met the criteria and that a specific analytical method was capable of satisfactory recovery. This filter device and analytical method became the basis of the analytical protocol. The protocol specifies requirements for:

a) laboratory procedures needed for DWI approval;
b) preparation of filter devices;
c) receipt of samples;
d) analysis and examination of samples;
e) reporting arrangements.

The regulations require that during normal continuous monitoring, samples are analysed within three days of the date on which the collection device was removed. Where there is an indication that the number of *Cryptosporidium* oocysts in the water may have increased, such as a significant rise in turbidity, the sample device should be changed as soon as possible and analysis completed by the following day.

The regulations also require that the analysis of sampling collection devices are only to be undertaken by approved laboratories. The DWI is currently undertaking inspections of laboratories to assess the accommodation, working procedures and personnel. Approval is granted when laboratories have demonstrated satisfactory provisions for analysis and reporting. Approved laboratories will be subject to unannounced audit to check the conduct of analysis.

5 VALIDATION OF NEW METHODS

The regulations require water companies to undertake sampling and analysis using only approved products and materials specified within the standard operating procedures. It is desirable to have a number of alternative approved products to prevent potential difficulties in meeting demand and encourage competitive pricing. Any modifications to parts of methods or alternative methods are required to be approved prior to use. The appropriate validation trials must be completed to establish performance. If the product or method demonstrates satisfactory performance in these trials DWI approval will be given. Specific guidance on the requirements for validation studies has been published to ensure that any new products or methods produce comparable performance to the existing approved products.

The validation process is in two stages:
a) Phase 1 - involves "in-house" evaluation of the new product or method to examine equipment suitability (where appropriate) and method performance parameters.
b) Phase 2 - is an inter-laboratory comparison to determine the precision and recovery of the new product or method under field conditions.

Various criteria are used to assess performance. For sampling any alternative sampling collection devices must be capable of processing water at a rate of 40 litres per hour for a period of 24 hours without blockage, leakage or breakthrough. In addition it must retain oocysts and allow elution to produce a recovery rate of at least 30% over the complete analytical process. Any new sampling collection device must be compatible with the requirements to demonstrate custody of the sample.

For any new or modified analytical methods validation is assessed against performance parameters for identity and selectivity, ruggedness, precision, bias (in terms of recovery) and usability. Again any new method must be capable of achieving a recovery rate at least 30% over the complete analytical process. New or modified methods are required to be assessed in parallel with the existing approved method. The performance of the new or modified method must be equal to or better than the approved method in these parallel trials. It has been observed that the approved method can produce recoveries consistently   above   50%   for   some   types   of   treated   water. The importance of ensuring that that the *Cryptosporidium* oocysts used for recovery tests are fully characterised has been emphasised in a recent workshop (11).

6 INTER LABORATORY PROFICIENCY SCHEME

Approved laboratories are required to participate in an Inter-laboratory proficiency scheme. The scheme will be organised by the Laboratory of the Government Chemist in conjunction with the Scottish Parasitology Diagnostic Laboratory. It will involve monthly distributions of three types of prepared test materials; sample filters, sample suspensions and microscope slides. The materials distributed may contain oocysts at various levels, oocyst like bodies and other interferents.

Each analyst within an approved laboratory, qualified to undertake microscopic examination, will be provided with an individual slide for examination. Each qualified analyst is required to participate in at least 10 of the 12 distributions per year.

The scheme will be administered by a Steering Board who will advise on the technical specification for the scheme and appropriate laboratory performance criteria. Performance indexes will be calculated for individual laboratories and in appropriate cases for individual analysts to indicate the performance of the laboratory or the analyst relative to other participants.

7 IMPLEMENTATION OF THE NEW REGULATIONS

The new regulations required water companies to complete their risk assessments by the end of August 1999. The DWI reviewed these assessments and confirmed the number of sites considered to be at significant risk by the end of October 1999. Water companies are required to submit a programme for installation of treatment and monitoring at these sites by the end of January 2000. This programme will be reviewed by the end of February by the DWI. Sampling equipment should be installed at the highest priority sites by the end of March with continuous monitoring to commence by the end of April. A phased programme of installation for the remaining sites will be agreed with DWI depending on the number of sites involved. Continuous monitoring at all sites will be required by the end of December 2000.

The DWI has notified water companies that continuous monitoring will not be required at sites where approved treatment capable of continuously removing or retaining particles greater than one micron has been installed. The DWI have issued an Information Letter (12) which lists the treatment processes currently approved and the requirements for operational monitoring to demonstrate continued membrane integrity.

8 REFERENCES

1. Cryptosporidium in Water Supplies; Third Report of the Group of Experts DETR/DoH HMSO
2. Cryptosporidium in Water Supplies; Report of the Group of Experts DETR/DoH HMSO
3. Cryptosporidium in Water Supplies; Second Report of the Group of Experts DETR/DoH HMSO
4. The Water Supply (Water Quality) (Amendment) Regulations 1999 Statutory Instrument No. 1524 HMSO
5. DWI Guidance on assessing risk from Cryptosporidium oocysts in treated water supplies.
6. DWI Standard Operating Protocol Part 1- Sampling and Transportation of Samples June 1999
7. DWI Standard Operating Protocol Part 2 - Laboratory and Analytical Procedures June 1999
8. DWI Standard Operating Protocol Part 3 - Validation of New Methods for Sampling and Analysis June 1999
9. DWI Standard Operating Protocol Part 4 - Requirements for the Inter-laboratory Proficiency Schemes June 1999
10. DWI Report on Continuous Sampling for *Cryptosporidium* in treated water supplies August 1999
11. DWI Report on a workshop on *Cryptosporidium* and water - Towards a standardised experimental design for viability and inactivation studies. August 1999
12. DWI Information Letter 16/1999

CRYPTOSPORIDIOSIS IN HEALTHY ADULT VOLUNTEERS

C. L. Chappell and P. C. Okhuysen

School of Public Health and Medical School
University of Texas Health Science Center
Houston, Texas 77030

1 INTRODUCTION

In the past, knowledge gained about human cryptosporidosis relied on outbreaks, cases of travelers diarrhea, and individual case reports. Most of what was known about the parasite came from animal studies and was assumed to apply to human infections. However, the animal models for *Cryptosporidium* consisted of neonatal animals with poorly developed immunity or adult animals that had been immunosuppressed. Further, the only models that developed a diarrheal illness were ungulates whose gastrointestinal tracts are distinctly different from humans. Due to the difficulties in finding a suitable human surrogate for *Cryptosporidium* infection, we instituted a series of studies in healthy adults volunteers. This work was deemed safe and ethical since the infection in healthy individuals is self-limited, the disease is not severe, and a large number of individuals are commonly exposed to this parasite through water or other modes of transmission. All of the volunteer studies have been approved by federal, state and private Institutional Review Boards. The studies have been designed to describe the natural history of the infection in humans, to establish the infectious dose (ID_{50}) of the organism and the illness attack rate, and to examine the immune response to the parasite. The data generated by these studies has added considerable knowledge regarding this disease in humans and has challenged some long held assumptions. This paper represents an overview of many of the findings of the past few years and poses directions for future research.

1.1 The Volunteer Model

The details of this model have been published previously.[1-5] Briefly, volunteers between the ages of 18 and 45 were recruited from the staff and student populations of the Texas Medical Center. These individuals were screened by ELISA for the presence or absence of anti-*Cryptosporidium* IgM and IgG. After a complete explanation of the study and information regarding the infection, the disease, transmission of the parasite, and other important information, volunteers who wished to enroll were given a written examination covering the important aspects of the infection. Only those who scored 100 were allowed to enroll in the study. Enrollment was completed by signing an informed consent document. Once enrolled, the volunteers were given a complete physical examination, including tests that would reveal any underlying illnesses or immunodeficiencies. Importantly, each person was tested for HIV and was documented to have immunoglobulin levels and T-cell subsets within normal limits. Any volunteers who lived in a household with children under two years of age or an elderly person or who had contact with any immunosuppressed person was excluded from the study.

Enrolled volunteers each received a single dose of *Cryptosporidium* oocysts delivered in a gelatin capsule. Volunteers were asked to withhold food for eight hours prior to and 90 minutes following oocyst ingestion. Volunteers were followed in the University Clinical Research Center (Herman Hospital) daily for 14 days and three times per week for an additional four weeks. During the initial two weeks, all stools were collected *in toto* and daily monitoring of vital signs and gastrointestinal symptoms was done. Also, an active household surveillance for diarrhea was carried out on a weekly basis for the period of the study. All stool samples were assayed by fluorescence microscopy using a commercial kit (Merifluor Diagnostics). Any diarrhea that occurred in volunteers or any household member during the study period was subjected to a work-up for the common enteric pathogens, including *Shigella, Salmonella, Campylobacter, Aeromonas, Plesiomonas,* and *Vibrio spp.* In addition, serum samples were collected before challenge and at days 5, 10, 30 and 45 after oocyst ingestion.

1.2 Oocyst Selection and Preparation

Three *Cryptosporidium* isolates were used in the volunteer studies. These isolates were selected for their ability to be amplified in neonatal calves, indicating that they were all genotype 2 oocysts. This assumption was confirmed by multilocus polymorphism analysis in the laboratory of Dr.Giovanni Widmer (Tufts University).[6] All three isolates yielded characteristic genotype 2 patterns when various markers (COWP, RNR, PolyT, ITS1, TRAP-C1, and □-tubulin) were used. The Iowa (provided by Dr. Charles Sterling, University of Arizona) and UCP (provided by Dr. Joseph Crabb, Immucell, Inc.) isolates were collected from naturally acquired infections in calves. Both isolates had been multiply passaged in calves in the laboratory and had retained their virulence for experimentally-infected calves and occasionally in humans accidentally infected in the laboratory. The third isolate (TAMU) was obtained from Dr. Karen Snowden (Texas A&M University) who collected a fecal sample from a veterinary student who had attended a necropsy of an infected foal. This isolate was passaged a small number of times at the University of Arizona before use in the volunteer studies.

Oocysts were amplified in calves in the laboratory of Dr. Charles R. Sterling (University of Arizona). Calf feces were collected into potassium dichromate. Oocyst purification followed a previously described protocol including graded sieving to remove particulates, discontinuous sucrose gradients, and isopycnic Percoll gradient separation.[7] An additional cesium chloride centrifugation was carried out to produce highly purified oocysts, which were then placed in a suspension of 2.5% potassium dichromate and shipped to Houston on ice overnight. Upon arrival at the Texas Medical Center, an aliquot of the oocyst suspension was removed under sterile conditions, and the potassium dichromate was replaced with phosphate buffered saline (PBS). These oocysts were then delivered to the clinical microbiology laboratory at Herman Hospital to be tested for any adventitious agents. Oocysts were stained (Gram, acid fast, fungal), cultured on eight standard bacterial media, including media for *Mycobacterium*, two standard fungal media, six standard cell lines for viral propagation, and subjected to an antigen capture technique for Rotavirus antigen. In addition, 10^6 oocysts were examined by electron microscopy. Inoculated media cultures were maintained for two weeks before final results were read and reported. All tests were required to be negative for the presence of any viable organisms (other than *C. parvum*) in order to qualify for use in the challenge studies.

Qualified oocysts were then washed free of potassium dichromate under sterile conditions and resuspended in PBS. Oocyst dilutions were counted with a hemacytometer, and adjustments were made to yield the desired number of oocysts. After the final adjustment was made, a minimum of six counts was done to ensure the accuracy of the inoculum suspension. A 10% coefficient of variation was achieved for inocula greater than 100 oocysts. For low dose inocula, the CV was higher. For example, when 10 oocysts were prepared, the mean and standard deviation was 9.2+/-3.9 with a 42.4% CV. The oocysts in a 10 µl volume were then instilled into a gelatin-filled capsule and delivered to the volunteers within one hour of preparation. Volunteers ingested the capsule with 250 ml of normal saline under the auspices of the investigators.

Oocysts were used within 6 weeks of calf production. Each batch of oocysts were examined for their viability and mouse infectivity at the University of Arizona (C.R. Sterling and M. Marshall). Viability was estimated by excystation rate using a previously described method.[8] All oocysts used in the studies had an excystation rate of 80% or greater at the time of the challenge. In addition, CD1 outbred neonatal mice were used to generate a dose response curve. Infectious dose estimates from these mice indicated that oocyst preparations were viable, infectious, and yielded repeatable values.

1.3 Definition of Infection and Illness

In a study, such as the one described herein, it is important to have carefully stated definitions so that the results can be interpreted in an appropriate manner and compared to other dose response studies.

Infection has been defined both parasitologically and clinically. The parasitological definition is the detection of oocysts in the feces. Detection, however, is limited, to fecal oocysts found 36 hours or more following challenge so that only replicating *Cryptosporidium* are included. Oocysts detected within the first 36 hours after challenge may represent only those ingested and not be indicative of infection. In some volunteers, particularly those known to have been previously exposed, the number of oocysts shed were dramatically decreased and often fell below the level of IFA detection (10,000 oocysts per ml). Since a number of these individuals had a diarrheal illness indistinguishable in onset, duration and symptomatology from that seen in the oocyst-positive volunteers, it was clear that IFA results were inadequate measures of infection in every case. (Subsequent flow cytometric studies have revealed low numbers of oocysts in the feces of IFA-negative volunteers.) Thus, a clinical definition of infection was developed that included any volunteer with a characteristic diarrheal illness (with or without demonstrated oocysts) that occurred within 30 days post challenge. Symptomatic volunteers were routinely studied for other enteric pathogens as described and were included only if these tests were negative. The published dose response studies using these definitions have included infectivity curves derived using each of these definitions. The ID_{50}'s quoted in the present paper represent the combined infectivity data from volunteers fitting either the parasitological and clinical definitions. *Diarrhea* has been defined in many different ways in the literature. We relied heavily on studies of traveler's diarrhea for developing the definitions used in our work and have included the character (formed versus unformed), frequency and volume of the stool. The definition of diarrhea used in the *Cryptosporidium* volunteer studies includes any of the following criteria: 1) production of 200 grams or more unformed (soft/loose/watery) stool per day, 2) 3 or more unformed stools in 8 hours, or 3) 4 or more unformed stools in 24 hours.

2 RELATIVE VIRULENCE OF *C. PARVUM* ISOLATES

Three *Cryptosporidium* (genotype 2) isolates (Iowa, UCP and TAMU) were used in the volunteer dose response studies. All volunteers enrolled in these studies were serologically negative (IgM and IgG) by ELISA.

2.1 Isolate Infectivity

The cumulative percent infection method of Reed and Muench was used to analyze the dose response data. Simple linear regression analysis was used to estimate the infectious dose (ID_{50}).[1,5] Two curves were generated for each isolate: one including only the volunteers in which oocysts were detected by IFA (confirmed infection), and another curve generated from volunteers exhibiting clinical cryptosporidiosis with or without IFA positivity (presumed infection). Table 1 shows the results of these analyses.

Table 1 *Infectious dose (ID_{50}) for three Cryptosporidium isolates in presumed and confirmed human infections.*

C. parvum isolate	Confirmed infection ID_{50}	Presumed infection ID_{50}
UCP	2788	1042
Iowa	75	87
TAMU	125	9

These findings showing logarithmic differences in *Cryptosporidium* infectivity for humans indicate the likelihood of wide variation in infectivity of oocysts circulating in communities. This variation in infectivity may help to explain the occurrence of outbreaks and suggests that certain isolates may be responsible for large outbreaks of human cryptosporidiosis. Further, these differences in infectivity occur within the population of genotype 2 oocysts that appear equal in current genetic analysis. This suggests that polymorphisms exist among these isolates that, when discovered, will lead to a laboratory indicator of potential infectivity. This advance will have enormous implications for the ability of water quality and public health officials to intervene in situations where community health is at risk.

Another interesting difference was seen in the ability of IFA to detect oocysts in the stools of these volunteers. With the Iowa isolate, 95% of volunteers with diarrhea had detectable oocysts compared to only 46% of volunteers receiving the UCP or TAMU isolates. In contrast, other measures of infectivity, such as the onset (4-7.7 days post challenge) and duration of oocyst shedding (3.3-8.4 days) were not statistically different among the isolates.

2.2 Clinical Outcome

The clinical outcomes and occurrence of a diarrheal illness in challenged volunteers was also monitored for each of the three *C. parvum* isolates.[5] The degree of difference in these data were not as dramatic as infectivity; however, the TAMU isolate yielding the lowest ID_{50} (9 oocysts) was also associated with the highest illness attack rate (86%). This attack rate was significantly greater ($p = <0.055$) than with the other two isolates, which were approximately 52-54% of exposed individuals. The similarity in illness attack rate between the Iowa and UCP isolates suggests that illness and infectivity are independent characteristics in *Cryptosporidium*. Interestingly, all three isolates were similar in mean incubation periods (5-9 days), mean duration of illness (1-9 days) and the mean numbers (6.7-8.2) and weights (936-1278 g) of unformed stools produced. However, it should be noted that, although not statistically significant, there was a tendency for greater stool volume in volunteers who had diarrhea from the TAMU isolate.

Based on the high serological positivity in human populations, it is thought that many *Cryptosporidium* infections may be asymptomatic or of a mild nature, which would not prompt the infected person to seek medical attention. Thus, it was of interest to compare among isolates the relative number of persons who had confirmed infection without diarrhea [unpublished data]. With the TAMU isolate, all individuals shedding oocysts also had diarrhea. Those receiving the UCP isolate had similar findings with only 1 (14%) person shedding oocysts without diarrhea. In contrast, the Iowa isolate showed that 6 (33%) subjects were asymptomatic or had mild symptoms. Thus, these findings indicate that asymptomatic or mildly symptomatic infections can occur in healthy hosts and may vary with the isolate. Since virulence factors for *Cryptosporidium* are not understood, the mechanism for these variations in clinical outcome remain a mystery.

3 EFFECT OF PRIOR EXPOSURE

Protection following prior exposure to *Cryptosporidium* has been examined in two ways—rechallenge of volunteers one year after initial challenge and challenge of volunteers who are serologically positive for anti-*C. parvum* IgG. Both of these studies utilized the Iowa isolate and monitored volunteers in the same manner as described for earlier studies.

3.1 Infectivity and Illness in Rechallenged Volunteers

All of the 29 volunteers who participated in the initial experimental infections with the Iowa isolate were invited to return for a second challenge one year later. Nineteen volunteers who were still located in Houston chose to do so. The results yielded several interesting observations.[3]

In these experiments, volunteers were all challenged with a single dose of 500 oocysts, which represented the ID_{80} for this isolate. Perhaps surprisingly, the same percentage of individuals (and often the same persons) developed a diarrheal illness after both challenges. Further, the characteristics of the illness were not different in incubation period (6.6 versus 8.2 days) or duration (2.8 versus 2.5 days). In contrast, the number of unformed stools passed was significantly lower after rechallenge (11.3 versus 8.6, p=<0.05) as well as the number of volunteers shedding oocysts (68% versus 16%, p=<0.005). Further, the rechallenged volunteers tended to have a delayed onset (9.8 versus 15 days) and shorter duration (6 versus 2.6 days) of oocyst shedding, although these values did not reach statistical significance.

3.2 Infectivity and Illness in Volunteers with Pre-existing Antibodies

Volunteers were selected for the presence of anti-*Cryptosporidium* serum IgG. These volunteers were given different doses of the Iowa isolate to compare infectivity (ID_{50}) in a naïve versus sensitized population of healthy individuals. The ID_{50} in volunteers with pre-existing antibody was approximately 20-fold higher than in serologically negative volunteers.[4] Further, the data indicated that the majority of individuals receiving low oocyst doses, as would be encountered in a water exposure, were protected against infection. Like the rechallenge studies, a significant percentage (46%) of volunteers with diarrhea had no fecal oocysts detectable with IFA. In those individuals shedding oocysts, the median onset (day 5, range day 4-19) and median duration (9 days, range 1-11 days) of shedding did not differ significantly.

The volunteers who developed diarrhea were among those receiving the highest oocyst dosages (10,000 oocysts or more). Again, those receiving low dose challenges were less likely to develop illness. Interestingly, the volunteers who did develop diarrhea had the same median incubation period (5 days, range 3-12 days) as the naïve hosts, but experienced a longer illness (6.5 days) with a greater number of unformed stools (10, range 3-35).

3.3 Diarrhea without Oocyst Shredding

A number of volunteers who developed a diarrheal illness had no detectable oocysts in their stools by IFA. This observation was especially prominent in volunteers who had been previously exposed to the parasite. The diarrhea in these individuals did not differ in the incubation time, duration, or any of the characteristic symptoms as compared to those who had confirmed infections. The cause of this phenomenon is not known, but several explanations exist. First, these diarrheal episodes may be the result of a hypersensitivity to the organism. In this scenerio, the exposure alone would provide a sufficient stimulus for the development of diarrhea without replication of the organism. However, with other antigens the hypersensitivity response typically requires 2-3 days or less to develop, a shorter time to onset than was observed with these volunteers. Secondly, an infection may have occurred with oocyst production below the level of IFA sensitivity (10,000 oocysts/ml). This possibility is enhanced by the finding of low numbers of oocysts when flow cytometry was used to assay the samples. Approximately 75% of the IFA-negative samples tested in this way were positive. Earlier studies indicated that flow cytometry of human stool samples has a detection limit of approximately 1,000 oocysts per ml, although fewer oocysts could be detected in some samples.[10] Interestingly, however, this left 25% of the samples negative by both methods. Thirdly, the asexual stages of infection could have occurred, but the immune response may have interrupted the cycle before the oocysts were formed. The factors necessary and sufficient for the development of diarrhea are not understood at present, but it is possible that parasite replication is needed to provide sufficient stimulus to initiate diarrhea. Likewise, the exact immune mechanisms governing clearance of the infection are also still unknown, although evidence suggests that $CD4^+$ T-cells and interferon gamma are essential components. Nevertheless, if the response is directed toward either the merozoite or gametocyte stages, oocyst production would be prevented. Further, it is also possible that a combination of these possibilities may be responsible for the lack of detectable oocysts.

4 ANTIBODY RESPONSE FOLLOWING *C. PARVUM* CHALLENGE

Cryptosporidium parvum infection remains localized just under the apical surface of the intestinal epithelium. The mucosal immune system is specifically designed to react to pathogens that infect epithelial cells and/or invade mucosal tissues. In contrast, systemic immunity is typically involved in the response to pathogens which invade deep tissues or circulate in the blood and lymphatic vessels. Thus, systemic and mucosal responses may be mutually exclusive. However, in cases where pathogens enter through the mucosa and invade other body tissues, both responses may be evident. It is also possible that continued or multiple exposures limited to the intestinal epithelium may eventually provide enough stimulus that systemic responses can be detected. This may take place via the organized (Peyer's patches) and/or diffuse intestinal lymphoid tissues (GALT). In this scenerio, serum responses could be indicators of reactivity occurring at a mucosal site. Since *C. parvum* is a non-invasive, non-systemic infection, the mucosal immune response is likely the first line of defense and the provider of protection from subsequent exposure.

4.1 Fecal IgA Extraction and Measurement

Stool samples collected from challenged volunteers were aliquoted and immediately frozen upon their arrival in the laboratory. For extraction, 2 mg thawed stool samples were diluted in 5.0 ml of PBS and homogenized. After sedimenting the heavier particulates, the sample was delipidated. The aqueous fraction was dialyzed (MWCO=50,000) overnight against PBS at 4°C to remove digestive enzymes and small molecules. The sample was then filtered through tandem membranes with 5 μm and 1 μm pore sizes. Extracts were frozen at –86°C prior to testing.

Samples were thawed and tested in batch. Specific IgA reactivity was detected in ELISA using disrupted oocysts as antigen (0.2 μg/well). Plates were blocked with 5% dry milk before extract was added in a 1:2 dilution to wells and incubated for 1 hour at 37°C. After washing, a peroxidase-conjugated anti-human IgA (heavy chain) was added to wells and incubated for 1 hour at 37°C. Unbound antibody was removed, and reactivity was visualized by adding ABTS substrate. Plates were read spectrophotometrically at 414 nm. For each volunteer, samples before and after challenge were all run in duplicate on the same plate. Data were expressed as absorbance values, and positive reaction was defined as any post-challenge absorbance value which exceeded 2 standard deviations above the prechallenge absorbance value.

4.2 Fecal IgA Reactivity in Challenged Volunteers

Fecal extracts were prepared from samples taken on multiple days after challenge with the Iowa isolate. All volunteers were serologically negative prior to challenge. An effort was made to select samples at four day intervals over the six week experiment when possible. Volunteers were categorized according to challenge outcome, and categories were compared for the number of individuals who had a specific fecal IgA response.

Table 2 *Specific fecal IgA reactivity to C. parvum antigens in volunteers challenged with the Iowa isolate.*

Challenge outcomes:	No. tested	No. of responders
Confirmed infection with diarrhea	10	7
Confirmed infection without diarrhea	4	2
No oocysts or illness	10	2

Of those with confirmed infections, extracts from multiple days were positive for specific fecal IgA. Specific reactivity was detected as early as day 5 post challenge and persisted in some for the duration of the study. In contrast, two individuals who were presumably uninfected each had only a single positive sample at day 34 and day 35, respectively. Statistical analysis revealed that infection (oocyst shedding) and challenge dose were significantly associated with specific fecal IgA (p=<0.035), while diarrhea alone was not [unpublished data]. Further analysis of specific fecal IgA showed no significant association with the duration of oocyst shedding or the intensity of the infection. Thus, specific fecal IgA was an indicator of active infection in volunteers shedding oocysts regardless of their illness outcome. The antibody appeared to remain high throughout the study period of 6 weeks, but since later time points were not examined, it is unclear how long the antibody persists. Another study examining serum IgA showed a decrease in the antibody over a short period of time.[11] If this proves to be the case with fecal IgA as well, the antibody may have value in identifying those persons who have an active or recent *Cryptosporidium* infection.

4.3 Serum Antibody Reactivity in Challenged Volunteers

Blood was collected before challenge and at days 5, 10, 30 and 45 post-challenge. Serum was separated, aliquoted and stored at –86°C prior to testing. ELISA was used to detect serum antibodies as previously described. Disrupted oocysts were used as antigen, and a biotinylated, monoclonal anti-human, isotype specific antibody was coupled with a peroxidase-conjugated streptavidin "reporter". Known positive and negative sera were included on each plate, and all control and test sera were run in triplicate. Reaction was visualized with ABTS substrate and read spectrophotometrically at 414 nm.

This assay was a modification of the method of Ungar et al., who employed a conjugated polyclonal anti-human antibody.[12] The modified method resulted in lower negative control absorbance values and higher signal to noise ratios. Typical negative control absorbance values are 0.25-0.3 and 0.1-0.15 for IgM and IgG, respectively. In comparison, positive control absorbance values are 0.8 or higher for each of the isotypes. Ideal positive controls yield an absorbance of 1.5 OD.

Post challenge sera (n=19) were collected from volunteers receiving a primary challenge and compared to sera collected from the same volunteers after second challenge one year later.[3] All of these sera were tested for IgM, IgG and IgA reactivity to Iowa isolate antigens. After the initial challenge 11 volunteers and 5 volunteers showed IgM and IgA responses, respectively. However, these responses were not associated with oocyst shedding or diarrheal illness. Further, serum IgA reactivity showed no correlation with individuals who had a demonstrable specific fecal IgA response. Surprisingly, no IgG response was detected in any of the volunteers following primary challenge, even those who were shedding oocysts and/or who developed diarrhea.

Volunteers were re-screened prior to receiving a second oocyst challenge one year after the initial challenge. None of the volunteers had demonstrable serum antibodies to *C. parvum* prior to the second challenge, even those who had IgM and/or IgA reactivities during the first study period. This observation supports the notion that both IgM and IgA responses are transitory and may only persist for a few weeks or months. Also, those who had an IgM or IgA response after primary challenge were not protected from infection or illness after a second oocyst challenge. After rechallenge, 6 volunteers had an IgM response, a decrease from that seen after primary challenge. Likewise, 6 volunteers had an IgA response, a number similar to that seen after primary exposure. Interestingly, 6 subjects showed significant increases in IgG after the second challenge when none had detectable antibody after the initial challenge. This finding suggests that a serum IgG response may require two or more challenges in order to yield a detectable response. We therefore assume that in population studies, the high percentage of seropositive individuals represent those who have been exposed to

Cryptosporidium more than one time. Thus, studies of communities following a *Cryptosporidium* outbreak may underestimate the number of persons exposed, especially if the pre-outbreak seropositives were in low proportion.

Persons who had detectable serum IgG to *Cryptosporidium* prior to exposure were studied for serological reactivity after oocyst challenge. Those with the highest IgG levels before challenge showed little to no increase in specific IgG after challenge. In contrast, subjects that had lower specific IgG levels prior to challenge showed a significant increase in specific IgG reactivity after challenge. These findings suggest that there may be a maximal level of reactivity that is not increased subsequent to another challenge; whereas, those who have lower IgG levels will be boosted with another oocyst challenge. How long these IgG antibodies remain in the serum is not known and await future studies.

5 SUMMARY OF THE VOLUNTEER STUDIES

5.1 Cryptosporidium parvum genotype 2 oocysts vary widely in their ability to cause infection and illness in healthy persons. Further, low oocyst numbers (<10) from certain isolates can result in significant infection and illness in a population.

5.2 Overall, about 24% of infected persons shed oocysts without having a diarrheal illness. These individuals would likely be responsible for secondary transmission, especially in non-household settings.

5.3 Persons who have serological evidence of prior infection (i.e. anti-*C. parvum* IgG) are relatively resistant to re-infection with low numbers of oocysts, as would be encountered in water sources. However, those who do become infected and develop diarrhea often experience a more severe illness than naïve persons.

5.4 Specific fecal IgA was associated with oocyst shedding in seronegative volunteers exposed to *Cryptosporidium* ostensibly for the first time. In contrast, none of these individuals developed serum IgG.

5.5 Serum IgG was detected in 33% of individuals exposed to *Cryptosporidium* oocysts a second time, indicating two or more exposures may be necessary to stimulate a detectable level of specific serum IgG.

5.6 Serum IgG levels in the "low detectable" range were increased after a subsequent exposure to *Cryptosporidium* oocysts.

Acknowledgements

The volunteer studies were supported, in part, by the U.S. Environmental Protection Agency (CR-819814 and CR-824759) and the National Institutes of Health General Clinical Research Center Grant, M01-RR-02558. We would also like to thank Dr. Herbert DuPont, Dr. Charles Sterling, and Mr. Walter Jakubowski for their helpful discussions and insightful comments, and Marilyn Marshall for supplying high quality oocysts. These studies could not have been done without the able assistance of Mrs. Madeline Jewell, Julie Rice, Nai-Hui Chiu, and the nursing staff at the Clinical Research Center at Hermann Hospital (Houston). In addition, we wish to thank the laboratory staff, especially N. Siytango-Johnson, H. Dang, C. Wang, M. Coletta, G. Nothdurft, and S. Baker for their excellent technical assistance.

References

1. H.L. DuPont, C.L. Chappell, C.R. Sterling, P.C. Okhuysen, J.B. Rose and W. Jakubowski, *N. Engl. J. Med.*, 1995, **30**,855.
2. C.L. Chappell, P.C. Okhuysen, C.R. Sterling and H.L. DuPont, *J. Infect. Dis.*, 1996, **173**,232.
3. P.C. Okhuysen, C.L. Chappell, C.R. Sterling, W. Jakubowski and H.L. DuPont, *Infect. Immun.*, 1998, **66**,441.
4. C.L. Chappell, P.C. Okhuysen, C.R. Sterling, C. Wang, W. Jakubowski and H.L. Dupont, *Am. J. Trop. Med. Hyg.*, 1999, **60**,157.
5. P.C. Okhuysen, C.L. Chappell, J.H. Crabb, C.R. Sterling and H.L. DuPont, *J. Infect. Dis.*, 1999, **180**,1275.
6. G. Widmer, L. Tchack, C.L. Chappell and S. Tzipori, *Appl. Environ. Microbiol.*, 1998, **64**,4477.
7. M.J. Arrowood and C.R. Sterling, *J. Parasitol.*, 1987, **73**,314.
8. D.B. Woodmansee, *J. Protozool.*, 1987, **34**,398.
9. L.J. Reed and H. Muench H, *Am. J. Hyg.*, 1938, **27**,493.
10. L.M.Valdez, H. Dang, P.C. Okhuysen and C.L. Chappell, *J. Clin. Microbiol.*, 1997, **35**,2013.
11. D.M. Moss, S.N. Bennett, M.J. Arrowood, M.R. Hurd, P.J. Lammie, S.P. Wahlquist and D.G. Addiss, *J. Eukaryotic Microbiol.*, 1994 **41**,52S.
12. B.L.P. Ungar, R. Soave, R. Fayer, T.E. Nash, *J. Infect. Dis.*, 1986, **153**,570.

TRIAL OF A METHOD FOR CONTINUOUS MONITORING OF THE CONCENTRATION OF *CRYPTOSPORIDIUM* OOCYSTS IN TREATED DRINKING WATER FOR REGULATORY PURPOSES

DP Casemore*, B Hoyle**, P Tynan*, Mark S Smith,*** with Members of PHLS project team.

* Public Health Laboratory Service Cryptosporidium Reference Unit, Rhyl, LL18 5UJ;
** Independent Water Treatment Consultant.
*** DETR Drinking Water Inspectorate.

1 INTRODUCTION

The role of Cryptosporidium as a waterborne pathogen, associated with treated potable water supplies, is now well established.[1] To date, more than 50 outbreaks have been reported in the literature from the developed countries, including the UK. The contribution to endemic (sporadic) infection is unknown and probably unquantifiable, although certain to occur. In the majority of outbreaks there has been some identifiable failure or reduction of water quality, especially changes, often brief, in turbidity which was thought to indicate an event permitting penetration of oocysts into the supply.[2] Such events are very difficult to detect using random single grab samples. Epidemiological investigation of outbreaks, and regulatory and operational needs in water treatment provided the opportunity to develop understanding and investigational methods for waterborne *Cryptosporidium*.[3-5] There was not, however, a consensus on the best methodologies, which were generally insensitive, variable, and insufficiently robust for regulatory purposes, leading to concern to improve the removal of oocysts from water and monitoring methodology.[5] Notwithstanding the methodological problems it was clear that oocysts are widely dispersed in the environment, in both surface and ground waters, and penetrate water treatment systems.[2] However, it is very difficult to interpret findings (numbers of oocysts/litre) with any certainty, particularly in relation to the significance of findings in terms of hazard (potential for harm) or risk (likelihood of harm) to the public health, i.e. health-based trigger values.[1] These are difficult to define or justify scientifically although they may be of value at a local level if part of a wider risk assessment based on local data on oocyst numbers, catchment assessment and infection prevalence. The alternative was to produce a water treatment based standard using continuous sampling.

1.1 Project background

In the late summer of 1995 an outbreak of cryptosporidiosis occurred in the south west of England, involving some 575 laboratory confirmed (stool-positive) cases among local residents and visitors.[6,7] A detailed epidemiological study suggested a link between the outbreak and consumption of water from one particular WTW. The company was charged with a Statutory offence but the Court ruled that the report of the OCT and analytical data were inadmissible as evidence. It is on the latter point that Government, through the DETR/DWI, propose to institute methods for the continuous monitoring of specified water supplies (DETR 1998), using formally taken samples. These samples should provide a continuous water process (water quality) parameter, which, in turn, could provide evidence, if subsequently required, that would be admissible in a Court of Law to a standard needed for a criminal prosecution - Police and Criminal Evidence (PACE) rules - under the terms of the Water Industry Act 1991. The DETR issued a Government consultation document, incorporating a draft Statutory Instrument (SI), to that end.[7]

1.1.1 The PHLS Role. In March 1998 the Public Health Laboratory Service (PHLS), which had been involved in methods development through the Cryptosporidium Reference Unit, was contracted by the Drinking Water Inspectorate (DWI) to develop an "off the shelf" analytical test procedure for the continuous monitoring of drinking water supplies for *Cryptosporidium* oocysts. The monitoring programme was intended for use in water treatment works assessed to be potentially at-risk of penetration of oocysts into treated water in distribution. The PHLS network of laboratories was able to provide logistical support and analytical facilities adjacent to proposed sampling sites. Logistical practicability, robustness and repeatability of the technique, practically and in statistical terms, for recovery of small numbers of oocysts, were considered to be paramount. Maximization of recovery was not required, as only a very short time scale was available for the project. Field trials were conducted, with full Water Companies' support, using water treatment works (WTWs) identified by the DWI. Sampling was by means of trial rigs that enabled more than one filter type to be evaluated in parallel, placed near to the point at which the Company takes samples for Regulatory compliance purposes.

1.1.2 Proposed Analytical Methodology. The proposed method comprised continuous filtration with daily sample filter changes, concentration, and detection/enumeration. Protocols were essentially similar to a United States EPA method, USEPA Method 1622,[8,9] developed as a result of a UK Government funded initiative, using commercially available technology. The US methodology comprised filtration of short run samples through cartridge membrane filters, immuno-magnetic bead separation (IMS), and immunofluorescence microscopy (IFAT). It was proposed to adapt this methodology for continuous sampling and to follow manufacturer's protocols for individual components of the system. The protocols were refined as experience and discussion within the steering group dictated.

2 THE MONITORING TRIALS

The principle objectives of the trial were as follows:
- Objective 1. To demonstrate the ability of the PHLS to carry out the work to the level of analytical and statistical precision and accuracy required, with appropriate quality control and quality assurance systems;
- Objective 2. To demonstrate the ability to produce 100 oocyst spikes with the necessary precision and accuracy, and with appropriate analytical quality assurance;
- Objective 3. To conduct a feasibility study of the logistics, and sampling and analytical methodology, over a sixty day continuous period, at three designated water treatment works, and to assess the reliability of the passage of ca 1,000L of treated drinking water through two designated trial filters over consecutive approximate 24 hr periods;

- Objective 4. To undertake quantitative recovery experiments, comprising replicate samples, using nominal 100 oocyst spikes, during the course of the 24 hr filtration of 1,000L of treated drinking water. The spiking trials were to be done at six different designated water sources, with the necessary precision and accuracy, and with appropriate analytical quality assurance and control. The rig used was to be in a closed system designed to prevent any possibility of oocysts inadvertently penetrating the water supply.

All stages were to be the subject of regular reports to DWI. It was anticipated that the trials would encounter a variety of problems to be addressed.

3 MATERIALS & METHODS

3.1 Sampling and Analytical Sites, and General Material and Methods

3.1.1 Water sampling sites. Water treatment works (WTWs) were identified by the DWI, to include a variety of different water sources and types. [10] Liaison between the DWI, the water companies and the analytical team, was via a designated member of the project team (BH) with expertise in water treatment.

3.1.2 Analytical laboratories. The PHLS sites chosen for the study included laboratories at Bristol, Chester, Leeds, and Rhyl, with further input from other units as required. The PHLS CRU at Rhyl PHL was responsible for co-ordinating the analytical aspects of the project, draft protocol production, ensuring adequate QA/QC, etc. The PHLS Environmental Surveillance Unit provided administrative support.

3.1.3 Technical procedures (SOPs) for the project. The methods outlined in the draft SOPs were adapted as required. Items of equipment materials and reagents used in the project were all readily available from approved suppliers.

3.1.4 Oocyst for Spiking. Batches of fresh, purified oocysts were supplied by PHLS Newcastle and Moredun Animal Health, Edinburgh. Use of live oocysts was specified and this was checked at CRU (DAPI/PI and excystation) using in-house protocols. Subsequently, it was agreed that field trials at WTWs would use freshly killed (heat-treated, 5 mins at $60^{0}C$) oocysts (non-viability confirmed by DAPI/PI and excystation).

3.1.5 Sampling. Samples for analysis were installed, collected and transported by project staff. All movements and procedures were documented and records maintained.

3.1.6 Sample Analysis. Manufacturers' protocols were followed for all analytical procedures, including filter elution, oocyst/deposit recovery and concentration, IMS (Dynal™), and IFAT (Cellabs Cryptocel IF™). In the latter case, however, as a result of in-house experience and data, monoclonal antibody staining time was extended to 60-90mins. DAPI staining of fixed oocysts was incorporated into the staining procedure as an aid to identification. Detailed protocols for the identification and enumeration of oocysts were adapted from those previously produced by CRU for the PHLS external quality assessment scheme for *Cryptosporidium* in water.[10] This included guidelines for the efficient and accurate microscopical scanning IFAT-stained samples on PTFE-coated slides. Although trials were to be conducted using purified, certified oocysts, outline criteria were defined to enable identification of putative oocysts in actual samples. These criteria depended upon the parallel use of a certified positive control for comparison. Arrangements were agreed with DWI and water companies on the action to be taken in the event of significant adventitious positive findings.

3.1.7 QA/QC. A full system of quality control was in place, including external monitoring of results; all work within the laboratory was logged and checked. The analyses included appropriate quality assurance (QA) and quality control (QC) samples, as specified in the SOPs The nature and frequency were as agreed with DWI, and included both qualitative and quantitative positive controls. An experienced statistician (Dr Hilary Tillett) applied statistical analysis to results. The laboratories involved were required to demonstrate initial and ongoing performance at a level acceptable to the DWI and their advisors, including the Laboratory of the Government Chemist.

3.1.8 Training. Initial training in the methods and procedures was by CRU to ensure that participants in the project were familiar with them. Videotapes provided for some of the commercial items were also used. Minimal technical supervision was given to sampling laboratories during trials as a test of the SOPs and robustness of the methods.

3.2 Objective 1

Background records, draft protocols, data, etc, were submitted and accepted by the DWI, following discussion at project steering group meetings.

3.3 Objective 2

3.3.1 Preparation of oocyst suspensions. Live oocysts not less than one month old were used with acceptable viability ($=>70\%$), satisfactory performance with specific monoclonal antibody-FITC- and IMS-conjugated reagents, and mono-dispersion. Approximately 10-fold ranging dilutions were prepared in test tubes, using graduated pipettes and micropipettes, to achieve the required stock suspension levels. At each stage, dilution steps were adjusted to achieve the desired concentration, depending on replicate counts. The methods for counting oocysts depended on the numbers present:

- Initial counts ($=>10^5$ oocysts/mL), in triplicate, in haemocytometers;
- For working (spiking) concentrations (ca10^2 oocysts, at 1/μL), stock suspensions were prepared estimated from stage one at 1,000 to 1,500 oocysts per mL. Counts for these lower numbers of oocysts were made by preparing replicates (x10) on PTFE well slides and staining with the approved IFAT/DAPI reagent;

QA/QC for counts: Results were analyzed statistically. Sources of potential error were defined and techniques and procedures used to minimize these.

3.3.2 Laboratory Spiking Procedure Checks. (i) 10 litre carboys of deionized water were seeded with ca100 oocysts ($+/-$ 20%), the oocysts were recovered by filtration through the selected filters, processed and enumerated, as described in the protocols supplied; (ii) The filters were incorporated into a full sampling rig connected to the mains tap in the laboratory. After running the supply for 10 minutes, the inlet connection (Hozelok™ fitting) to the filter was disconnected and an aliquot of the oocyst suspension containing approximately 100 oocysts was injected into the inlet to the filter. The inlet connection was then reconnected and 10 litres or 100 litres (metered) of tap water was run through the rig. The oocysts were recovered and enumerated, as described in the draft protocols; (iii) after the initial spiking tests using 10L and 100L volumes, the procedure was repeated but with approximately 1000 litres of tap water filtered. Unspiked negative control samples were run in parallel for QC purposes.

3.4 Objective 3: 60 day Logistics Trials

Two types of filters were used in the trials, the Gelman Envirochek Capsule™, a pleated membrane cartridge filter with a nominal pore size of 1.0 µm, and the Genera Crypto-D-tec™ (subsequently re-named Filta-Max™), an in-depth filter consisting of multiple layers of compressed reticulated open cell foam. The Envirochek filter is quoted as having a maximum operating differential pressure (headloss) of 2.1 bar; no operating pressure restrictions have been placed on the D-tec™ filter although the housing has been reported as being successfully tested to 14 bars.

   3.4.1 Trial sites. The sites selected for the 60 day trials were specified by DWI and covered a range of water qualities and treatments.

   3.4.2 Trial filters. The two types of filters were installed at each site running off a single sampling tap with a Y-connector dividing flow between the two filters. The sampling lines were made up using flexible hose and Hozelok™ connectors to allow rapid change of the filters. A flow restrictor (Meric™, usually with a nominal 1 litre per minute flow when operated at pressures between 1 and 10 bar) was installed after each filter, after which the flow passed through Kent™ flow meters that give the accumulated flow through each filter.

   3.4.3 Sample Analysis. Protocols were followed for analytical procedures, as described above. All results were logged and submitted to DWI at regular intervals, along with progress reports, indicating any problems experienced.

3.5 Pressure Monitoring (Filter Integrity) Trials

These trials involving monitoring the headloss development across the Envirochek™ and D-tec™ filters. The trials were made using distributed water from a domestic supply (BH). Four test runs were made, using Envirochek™ filters and three runs using D-tec™ filters.

3.6 Objective 4: On-site Spiking trials

Quantitative recovery experiments, comprising ten replicate 100 oocyst spikes of 1,000 litres of treated drinking water, were conducted at six different water sources using Genera Crypt-dtec™ filters only (see below). The experimental rig was as described above. Spikes were also run at different timed intervals to assess the impact of this on recoveries.

   3.6.1 Site Trials - Sampling. These trials were carried out at one of the works used in the 60-day trials and a further 5 treatment works. All of the works were treating surface derived sources of varying quality. The treatment employed at the six works was also different and covered a wide range of processes including slow sand filtration, chemical coagulation, rapid gravity filtration, direct pressure filtration, dissolved air flotation, pH correction (lime or sodium hydroxide), chlorination and ozonization.

   The trials were carried out using a break tank and non-return valve to ensure that no contamination of the supply could occur through reverse flow. A small centrifuge pump was used to provide flow in each trial and water meters and flow controllers were fitted to each sample line. Suspensions of oocysts used in the trials were all heat deactivated to ensure that the oocysts were non-viable. Based on the results of stage three (logistics) trials, two D-tec™ filters were run in parallel in all of these tests together with a negative control filter to check that there were no oocysts present in the supply water.

Backup filters were fitted downstream of the sample filters to capture any oocysts that penetrated the sample filters. The backup filters were usually left in place for the five days of the spiking trial unless there was a noticeable fall off in the volume of water filtered during the previous 24-hour period. If such a fall off in filtered volume occurred then the backup filter was changed at the same time as the sample filter.

The equipment was assembled and attached to the sample tap, following flushing, and the tap was opened to allow the break tank to fill. The pump was then switched on and the flow through the assembled rig checked to ensure that the flow through each line was satisfactory (ca 1 litre per minute). The supply was then turned off and the meter reading noted. The sample hose lines were disconnected immediately upstream of each of the two sample filters and the hose end leading to the sample filter held in an upright position. Using a vortex mixer the oocyst working suspension was mixed for two minutes. The required volume to give a spiking dose of approximately 100 oocysts was injected using a micropipette directly into the open end of the pipe leading to a sample filter. The hosepipe was then reconnected, and the process was then repeated for the second sample filter. The pump was then switched on and the time of starting up noted. At the end of the sampling period the pump was switched off and time noted, the two sample filters units were removed for analysis together with negative control filter unit, and new filter units fitted and the whole procedure repeated for a further four 24-hour periods. At the end of each site trial the backup filters were also removed for analysis.

3.6.2 *Site trials - Analysis.* Samples were analysed, logged, and reported to DWI as described above.

4 RESULTS & DISCUSSION

4.1 Preparation of oocyst suspensions

Statistical analyses of the results from the first three suspensions (Sets A, B and C) confirmed that the data demonstrated successful preparation of working oocyst suspensions, generally within statistically acceptable limits of 1 oocyst/µl in 25µl aliquots. Set B (Table 1) gave excellent counts of oocysts that varied around the mean value with random variation over time and between replicates. The variance of the 100 counts was slightly but not significantly higher than the mean count. In perfect random distribution the variance equals the mean. Set C data exhibited over-dispersion and changes in mean value between some days. There was a problem with clumping.

Table 1 *Enumeration of Suspension B*

Date	1	2	3	4	5	6	7	8	9	10	Mean	Std
06/07/98	31	22	18	26	29	22	22	32	30	27	25.90	4.70
07/07/98	29	28	23	21	21	25	18	15	25	25	23.00	4.35
08/07/98	22	21	22	24	19	31	20	17	32	19	22.70	5.03
09/07/98	19	23	28	14	18	26	17	15	25	23	20.80	4.85
10/07/98	22	28	21	16	23	31	32	27	18	20	23.80	5.45
11/07/98	27	16	23	19	25	22	25	29	28	27	24.10	4.15
13/07/98	20	10	23	24	27	16	18	20	20	26	20.40	5.04
14/07/98	21	19	15	29	18	10	18	23	25	23	20.10	5.36
15/07/98	32	14	26	22	24	24	25	20	19	18	22.40	4.99
16/07/98	36	24	30	21	27	29	20	15	20	20	24.20	6.25

4.2 Preliminary laboratory spiking trials

The results of the preliminary spiking trials using the Gelman Envirochek™ showed significant oocyst losses as the volume filtered increased (23% from 10L, 18% for 100L, 1% for 1000L). With the Genera D-tec™ filter there was statistically negligible differences in average recovery rates when filtering 10, 100 or 1000 litres of water (24% for 10 litres, 21% for 100 litres and 22% for 1,000 litres). These preliminary recovery values, although less than expected, were considered adequate to proceed to field trials.

4.3 60-Day Logistics Trials

After some initial logistical problems, it was found that sampling, transport and analysis for the three sites was practicable. Problems were found with various parts of the sampling process and system, which were resolved by discussion and/or method modification. In some cases, problems led to chemical mass balance analysis for the WTWs (BH) and to modification of treatment.

4.3.1 Daily Sampling. The flow volumes initially were generally on the low side of that desired through the Envirochek™ but adequate or better through the D-tec™ filter. Daily flow values for site 3 showed that the flow from this works was distributed by gravity and therefore the pressure available at the sampling point was too low (approximately 500 to 1000 millibars) and a pump had to be installed. After installation the centrifuged deposit volume recovered from the Envirochek™ filter averaged only 0.04 millilitres per cubic metre of water filtered, whilst the volume recovered from the D-tec™ filter averaged 1.22 millilitres. When the centrifuged deposit volumes recovered from all of the filters were analysed, the average recovered from the Envirochek™ filter were approximately half that for the D-tec™ filter. The reason for the small amount of deposits recovered from the Envirochek™ filter was found to be due to damaged caused to the filter through high headloss development (see 4.4.).

The hose connections supplied on both types of filters were unsatisfactory and resulted in leaks occurring in routine sampling. However, those on the D-tec™ filter could readily be replaced with more suitable fittings.

4.3.2 Sample Analysis. Sample analysis results were, logged and reported to DWI as described above. Confirmed adventitious oocysts were not detected during the trial period. Some D-tec™ filters failed to expand fully during elution. This was believed to be due to excess water treatment chemicals in the treated water (see above). This did not appear to significantly affect recoveries during spiking trials (see below).

4.4 Pressure Monitoring Trials

It was a requirement of the contract that filtration should, so far as possible, be representative of flows throughout the 24hr period. In addition, some of the problems encountered, especially the reduced volume of deposit from the Envirochek™ filter, needed to be investigated. A separate trial was set up (by BH) using a domestic water supply, with logging of head-loss data.

The headloss measured across the Envirochek™ filter was generally too high and, for example, on one occasion was 5590 millibars at the end of a sampling run. This headloss was far in excess of the quoted maximum operating headloss of 2100 millibars, and led to collapse of the membrane support mesh (Figure 1). The headloss measured across the D-tec™ filter was 1650 millibars on the same date. Turbidity of the treated

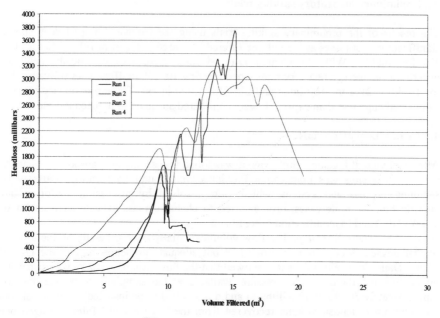

Figure 1 *Gelman Envirochek^{TM} filter variation in headloss with volume filtered*

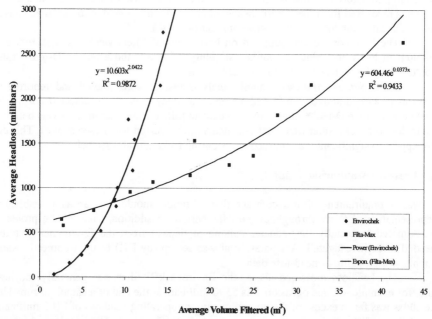

Figure 2 *Envirochek^{TM} vs. Filta-Max^{TM} filters comparison of headloss buildup*

water ranged from less than 2 to 16 FTU, and averaged 0.7 FTU.

Results of the pressure challenge runs were logged. The reason for the rapid fall-off in headloss during some runs with this filter was initially unclear as the filter appeared to be sound with no visible signs of distress. For the Envirochek™ filters it was noticeable in some cases, however, that the membrane support mesh started to collapse once the headloss exceeded the maximum operating pressure. Although reforming of the pleating followed collapses early in the run (<=24hrs), it was noted that the membrane support mesh had completely collapsed by the end of the trial runs (>24hrs). Results from the four runs using Envirochek filters and the three runs using D-tec™ filters were grouped in headloss ranges and then averaged. The average results for each type of filter was plotted and a best-fit curve (linear, polynomial, logarithmic, power and exponential fits were tried) was then applied to each set of data (Figure 2).[10]

From the results of the laboratory spiking and objective 3 field trials it was apparent that the Envirochek™ oocyst recovery rates and filter deposit volumes were unsatisfactory when filtering volumes of ca1,000 litres. The 60-day trials indicated that the D-tec™ filter could filter 1,000 litres providing the inlet pressure was satisfactory, whereas the Envirochek™ filter was often unable to cope with filtering this volume of treated water without exceeding the design headloss. Recovered deposit volumes from the Envirochek™ filter were also significantly lower than those for the D-tec™ filter due to loss during collapse. The pressure monitoring trials confirmed that the D-tec™ filter was able to filter greater volumes of treated water than the Envirochek™ filter with more even flow rates throughout the 24hrs sampling period. Leakage from the connecting spigot was sometimes a problem but was considered more readily remediable with the D-tec™ filter. For all of these reasons it was agreed that the rest of the trials should proceed using only the D-tec™ filter.

4.5 On-site Spiking Trials

Table 2 *Summary of site spiking trials*

Treatment Works	Recovery (mean)	(std. Dev.)	Slippage (mean)	(std. Dev.)
Site A	42.13%	8.84%	0.67%	0.39%
Site B	36.73%	10.06%	0.67%	0.66%
Site C	22.83%	4.72%	2.08%	0.09%
Site D	27.78%	5.70%	0.56%	0.12%
Site E	29.15%	11.43%	0.44%	0.09%
Site F	42.21%	11.75%	0.09%	0.09%
Site C (2) - 0 hours	33.97%	7.56%	1.41%	0.51%
Site C (2) - 6 hours	42.02%	6.99%	2.40%	4.18%
Site C (2) - 18 hours	43.71%	13.56%	4.65%	5.90%
Site C (2) - 23.75 hours	47.94%	9.93%	1.27%	0.87%
Overall	35.69%	11.43%	1.15%	1.88%

The results from the trials at the six sites are summarised in Table 2. The recoveries at the six sites varied between 23% and 42%, and averaged 33.5% (35.7% including timing trials). It is thought the variation shown in recovery is due to inherent variation in the method, especially using such small numbers of oocysts, rather than due to differences in water quality characteristics at the six sites. This tends to be substantiated by the additional trials carried out at Site C when on a second trial spiking the water at the commencement of the run the average recovery was found to be 34% (cf. 23% on the initial trial) although this extra run was excluded from the calculations.

Further trials at Site C were carried out to check whether spiking the filters at the beginning, middle, or towards the end of run made a significant difference to the recovery rates (Table 2). The results suggest that there is a small but statistically insignificant improvement in oocyst recovery rates the later into the sample run the spike is applied. On two occasions relatively large slippage (9.9% and 14.8%) of oocysts occurred. No reason was identified for the slippage.

4.6 Recovery Rates from Various Stages of Washing & Analysis

An assumption was made at the outset, based on in-house and published data, that each of the four stages of the test procedure could be expected to achieve 80% efficiency at best, yielding an expected overall recovery of about 40%. Tests were therefore carried out to check the recovery rates of oocysts from various stages of the procedure. The results show an overall average oocyst recovery of 64%; 84% from spiking at the primary concentration stage; a similar recovery (83%) from spiking at the secondary concentrate stage; spiking at the IMS stage the recovery rose to 92%; the greatest loses (44%) were at the washing stage and were believed to result from excess centrifugation steps and/or inadequate speed and/or time of centrifugation. From this work and recovery rates achieved in the field spiking trials, average losses of oocysts through the various stages of the sampling and analytical procedures were calculated and recommendations made to modify the protocol. Subsequent studies (J. Watkins, D Casemore, unpublished data) showed that overall recoveries could readily exceed 60-80%.

5 CONCLUSIONS

The trials confirmed the logistic and analytical feasibility of the recovery of 100 oocysts in 1000L taken over a 22-24hr period, using the D-tec™ filter, and the robustness of the protocols developed.[10] The trials highlighted the need to optimize water pressure, with supplementary pumps if necessary. The trial rigs, and analytical and other findings, provided critical information for the design of the substantive engineered sampling units. Occasional failure of the D-tec™ filter to fully re-expand did not significantly affect recoveries. The protocols were suitable for developing versions compliant with the PACE rules and the DWI protocols and the new Statutory Regulation has been issued.[12,13]

6 REFERENCES

1. Meinhardt P, Casemore DP, Epidemiological aspects of human cryptosporidiosis and the role of waterborne transmission. *Epidemiologic Reviews* 1996; **18**: 118-136.

2. Bouchier I (Chairman). Cryptosporidium in water supplies. Third report of the Group of Experts to: Dept of Environment, Transport and the Regions & Department of Health. 1998.

3. Casemore DP. Problems with sampling and examination for *Cryptosporidium*. In: Eds Dawson A, Lloyd A. Workshop on *Cryptosporidium* in water supplies. HMSO London, 1993, Pp 11-17.

4. Anon. Isolation and identification of Giardia cysts, *Cryptosporidium* oocysts and free-living pathogenic amoebae in water and associated samples. HMSO London 1990.

5. Casemore DP. The problem with protozoan parasites. In: eds WB Betts et al. *Protozoan parasites and water*. Royal Society of Chemistry, Cambridge, 1995, Pp 10-18.

6. Waite WM. DWI Assessment of water supply and associated matters in relation to the incidence of cryptosporidiosis in Torbay in August and September 1995. DETR/Welsh Office. 1997.

7. Anon. Preventing Cryptosporidium getting into public drinking water supplies. Consultation paper. DETR/Welsh Office 1998.

8. Anon. EPA Method 1622: *Cryptosporidium* in water by filtration/IMS/FA (December 1997 draft). United States Environmental Protection Agency, Office of Water, Washington DC.

9. Schaub S, Fox J, Regli S, et al. EPA protozoa method criteria document for guidance in development of analytical methods (Draft September 1977). United States Environmental Protection Agency, Office of Water, Washington DC.

10. Anon. Continuous sampling for Cryptosporidium in treated water supplies. Report to the drinking Water Inspectorate on studies into the feasibility of continuous sampling for Cryptosporidium oocysts in treated drinking water supplies. DETR, National Assembly for Wales, 1999.

11. Lightfoot NF, Place B, Richardson IR, Casemore DP, Tynan PJ. External Quality Assessment Scheme for *Cryptosporidium* in water. In: eds WB Betts et al. *Protozoan parasites and water*. Royal Society of Chemistry, Cambridge, 1995, Pp 164-7.

12. Anon. Statutory Intruments 1999 No 1524. Water Industry, England and Wales. The Water Supply (Water Quality) (Amendment) Regulations 1999.

13. Anon. Standard Operating protocol for the monitoring of *Cryptosporidium* oocysts in treated water supplies to satisfy Water Supply (Water Quality) (Amendment) Regulations 1999, SI no 1524. Part 2 - Laboratory and Analytical Procedures. DETR/Welsh Office June 1999.

7 ACKNOWLEDGEMENTS

The authors were supported by various people but particularly wish to acknowledge the help of A Bryden, P Rushden, C Woodward and S Eggleston (PHLS) and their colleagues for sampling and analytical work, Hilary Tillett for statistical support, and G Nichols for administrative support; North-West Water, Yorkshire Water and Bristol Water Company provided access to trial sampling sites.

This work was funded by the Department of the Environment Transport and the Regions and was managed by the Drinking Water Inspectorate. Any views expressed are those of the authors and not necessarily of their respective organizations.

A DIELECTROPHORESIS SYSTEM FOR RAPID ANALYSIS OF *CRYPTOSPORIDIUM PARVUM*

A.P. Brown and W.B. Betts

Cell Analysis Ltd
Institute For Applied Biology
University of York
York YO10 5YW

1 INTRODUCTION

The currently accepted U.K. Regulatory method for *Cryptosporidium* is based on the continuous filtration of a 1000 litre water sample over a period of 24±2 hours using a compressed foam filter.[1] Subsequent to sample collection the filter is analysed for *Cryptosporidium* content by elution, centrifugation, purification by immunomagnetic separation and finally microscopic examination using fluorescent monoclonal antibodies. The time delay incurred before results of *Cryptosporidium* analysis are available indicate that a more rapid method is required; this would have enormous public health advantages and potential labour and cost benefits. Dielectrophoretic methods have the potential to offer such advantages.

Dielectrophoresis (DEP) is the term used to describe the polarisation and associated motion induced in particles or cells by a non-uniform electric field.[2] The phenomenon arises from the difference in the magnitude of the force experienced by the electrical charges within an unbalanced dipole, induced when a non-uniform electric field is applied. The dielectrophoretic force on the cell, or oocyst is greater in regions where field non-uniformity (or gradient) is greatest. If the particle, is more easily polarised by the electric field than the medium in which it is suspended, it will experience an attraction to regions where the electric field has most non-uniformity, such as an electrode. This movement is known as positive dielectrophoresis and can result in collection of large numbers of particles upon electrodes in real time. Since the force is related to the frequency of the electric field, due to the dielectric characteristics of the cells and medium, the amount of collection upon electrodes varies according to the frequency of the applied electric field. Examination of the amount of collection upon electrodes over a frequency range can enable production of a spectrum which is distinct, and characteristic for a particular cell type. Frequency spectra have been produced for many types of prokaryotic and eukaryotic organisms, including *Cryptosporidium parvum*, and also for abiotic colloidal particles such as polystyrene latex.[3,4]

Cell Analysis Ltd, at the University of York, use a semi-automated system in order to selectively enrich, separate and characterise cells by exploiting their dielectrophoretic response to an alternating non-uniform electric field. Microelectrode chambers are used to manipulate cells and an image analysis microscopical method quantifies collection to produce spectra. The system is a patented recirculating system, enabling measurements of DEP collection to be made very quickly. Spectra can be produced within a few minutes using this technique.

DEP spectra obtained to date indicate that there are many differences in the frequency responses of oocysts that could be exploited for both identification and separation purposes. Furthermore, the speed of DEP analysis would be of great benefit to improvements in time-consuming and labour intensive testing. The feasibility for rapid DEP separation of viable and non-viable *C. parvum* has been described previously.[5-7]

However, to date, the majority of dielectrophoretic analyses of *Cryptosporidium* have been performed using model suspensions in deionised water media. The preliminary study presented here was conducted on end-of-pipe tap water samples, following standard methods of preparation used previously by water testing laboratories.

2 MATERIALS AND METHODS

2.1 Oocysts

Cryptosporidium parvum oocysts were provided by Yorkshire Water plc, as a suspension in PBS (10^8 oocysts/ml).

For production of a reference frequency spectrum, a sample of oocyst stock was diluted in deionised water, recirculated through the microelectrode chamber and the number of oocysts collected by DEP over a frequency range was counted. The spectrum is shown in Figure 1.

2.2 Water Samples

Two 10 L grab samples were taken in clean, sterile containers. One of the samples was spiked with, a known concentration of oocysts. The other 10 L tap water sample was used as an unseeded control sample. These were filtered through 1.2 µm pore cellulose acetate membrane filters, added to stomacher bags and a 200 ml volume of Tween distilled water (TDW; 0.1 % Tween 80 in distilled water) then added to each. The membranes were massaged using fingers for 1 min and the suspensions centrifuged at 1500 x g for 10 min. Each pellet was then resuspended in a final volume of 2 ml TDW ready for processing by DEP. The TDW was found to have a very low conductivity (5.00 µS/cm), which could easily be used for positive DEP experiments.

The suspensions of filtered material were passed through the DEP analysis system without any further clean-up. Both spiked and control samples were analysed individually by application of electric field frequencies between 1 kHz and 50 MHz with duration of each field application being 20-30 seconds. The amount of collection at each frequency was enumerated by image analysis and replicate spectra were obtained for each sample. The spectra obtained are shown in Figure 2.

3 RESULTS

The reference spectrum (Figure 1) is similar to those obtained previously for *Cryptosporidium* using this system at York.[4,5] This spectrum shows a flat region up to 100 kHz, whereafter the collection begins to rise dramatically. This reaches a peak height just after 1 MHz frequency and subsequently begins to fall steadily again.

The seeded sample produced a spectrum (Figure 2, ■) which was identical in form to that of the reference sample, showing the rise in collection after 100 kHz with a peak at 1 MHz.

The spectrum of the control unseeded sample (Figure 2, ▲) was relatively flat over the whole frequency range, producing collection levels of less than 20 particles (per field of view). Due to the dissimilarities in spectrum between the spiked sample and the unseeded sample, it is considered that the spectrum of the seeded sample was due mainly to the collection of oocysts rather than any of the other material. This was confirmed by the shape of the spectrum, which was identical to the reference spectrum of known *Cryptosporidium*.

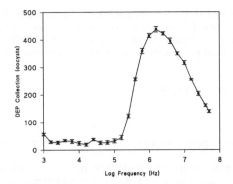

Figure 1 *Reference DEP frequency spectrum of Cryptosporidium parvum oocyst suspension in deionised water (–■–). Parameters: 16 V, 20 s field duration, 5 rpm collection speed, 2 rpm release speed, 5.025 x 10^6 oocysts/ml. DEP collection reflects numbers of oocysts per field of view.*

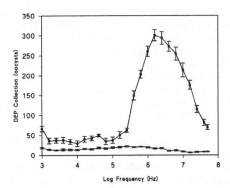

Figure 2 *DEP spectra of tap water spiked with 4.64 x 10^6 oocysts (–■–), and a control tap water sample (–▲–). Parameters: 16 V, 20 s field duration, 5 rpm collection speed, 2 rpm release speed, 4.45 µS/cm. DEP collection reflects numbers of oocysts per field of view.*

4 DISCUSSION

This set of preliminary experiments demonstrated the potential of dielectrophoresis to collect *Cryptosporidium* oocysts from seeded tap water samples on microelectrodes, and to potentially discriminate between oocysts and the normal suspended solids present in real world water samples.

High levels of oocysts were used to spike the water samples (4-5 x 10^6 oocysts), which were much greater than might normally be found in potable water samples. Further research is being undertaken to enable examination of lower concentrations of oocysts. In its current form, this DEP system may be able to examine oocyst suspensions of much lower concentration (several orders of magnitude less) but improvements in sensitivity to reach current regulatory requirements for *Cryptosporidium* analysis would undoubtedly be necessary.

The non-*Cryptosporidium* debris (particularly bacteria and algae) filtered out from the water samples will also undergo dielectrophoretic collection, and due to their presence in greater numbers than normal *Cryptosporidium* levels, it is anticipated that their DEP collection may be greater than that found with *Cryptosporidium*. Though bacterial cells can collect very well by DEP, their contribution to spectra was quite small in this case, reflecting the optimisation of the image analysis for the detection of larger particles.

To reduce non-specific DEP collection, the operating conditions could be optimised to use those which will favour the collection of *Cryptosporidium*. By manipulation of frequencies, conductivities and other operating conditions, it should be possible to select parameters under which *Cryptosporidium* oocysts will collect on the microelectrodes by dielectrophoretic force but which will not collect other organisms.

DEP is a rapid selective technique which could improve time consuming techniques such as microscopy and viability determination, or offer an alternative separation procedure to immunomagnetic beads for sample clean up. It has the ability to discriminate between viable and non-viable oocysts on the basis of changes to frequency spectra,[4,7] and may also be used as a system for confirmatory identification of oocysts on the basis of their frequency response without the need for fluorescent antibody labelling even with single oocysts.

Acknowledgements

We gratefully acknowledge the support of Dr Gary O'Neill (Yorkshire Water plc) for supply of oocysts, Mrs Irene Watson for assistance with preparatory methods and Mr Keith Milner for dielectrophoresis microelectrode manufacture.

References

1. Standard operating protocol for the monitoring of *Cryptosporidium* oocysts in treated water supplies to satisfy Water Supply (Water Quality)(Amendment) Regulations 1999, S.I. No. 1524. Drinking Water Inspectorate. Part 2-Laboratory and Analytical Procedures.

2. H.A. Pohl, Dielectrophoresis: The Behaviour of Matter in Non-Uniform Electric Fields, Cambridge University Press, Cambridge,1978.

3. A.P. Brown and W.B. Betts, *J. Appl. Microbiol. Symp. Suppl.*, 1999, **85**, 201S.

4. G.P. Archer, W.B. Betts and T. Haigh, *Microbios*, 1993, **73**, 165.

5. C.M. Quinn, G.P. Archer, W.B. Betts and J.G. O'Neill, in *Protozoan Parasites and Water*, ed. W.B. Betts, D. Casemore, C. Fricker, H. Smith and J. Watkins, Royal Society of Chemistry, Cambridge, 1995, p. 125.

6. H.V. Smith, L.J. Robertson and A.J. Campbell, *Europ. Microbiol.*, 1993, Jan-Feb, 22.

7. C.M. Quinn, G.P. Archer, W.B. Betts and J.G. O'Neill, *Lett. Appl. Microbiol.*, 1996, **22**, 224.

A REVIEW OF METHODS FOR ASSESSING THE INFECTIVITY OF CRYPTOSPORIDIUM PARVUM USING IN-VITRO CELL CULTURE ·

P. A. Rochelle and R. De Leon

Water Quality Laboratory
Metropolitan Water District of Southern California
La Verne, California 90266

1 INTRODUCTION

Many methods exist for detecting *Cryptosporidium parvum* oocysts in environmental and finished waters. Some of these methods are approved or mandated by environmental and regulatory agencies, while others are primarily research tools. These methods have varying recovery efficiencies and reproducibility characteristics. In addition, there is an ongoing effort to develop new methods and improve existing techniques. However, to fully assess the public health significance of waterborne *C. parvum*, the water industry needs the capability to determine whether waterborne oocysts are infectious. A major application of such methods would be measuring the efficacy of disinfection and treatment strategies. Ultimately, infectivity information will allow a more informed industry response on occasions when oocysts are detected in either source or finished waters.

The three methods available for investigating infectivity of parasites are human volunteers, animal models, and in-vitro cell culture. Human volunteer and animal infectivity studies, while useful, are not practical for use on a routine basis by the water industry. Also, standard animal models (e.g., calves and neonatal mice) do not appear to support infection with all genotypes of *C. parvum*. In contrast, cell cultures have been shown to support growth of both genotype 1 and genotype 2 *C. parvum* (P. Rochelle, unpublished; J. Mead, pers. comm.). Therefore, cell culture- based infectivity assays are a promising alternative to human and animal-based techniques. Apart from determining the infectivity of environmental oocysts, cell culture-based methods have also been used in attempts to propagate *C. parvum*, to study the biology and life cycle of the parasite, and to evaluate the efficacy of potential anti-cryptosporidial chemotherapeutic agents.

This paper reviews the methods used for in-vitro cell culture of *C. parvum*, describes techniques used to measure the amount of infection, and discusses applications and limitations of cell culture methods.

2 CELL LINES AND CULTURE CONDITIONS

At least 20 cell lines have been shown to support *C. parvum* infection (Table 1). Infection in cell culture is defined as the development of one or more intracellular stages of the *C. parvum* life cycle, following inoculation with oocysts or excysted sporozoites. Such stages include trophozoites, meronts, macrogametes, and micro-gametocytes (Figure 1). Complete development of *C. parvum* is defined as the de novo production of oocysts as a result of the parasite undergoing its full life cycle. Complete in-vitro development in cell culture was first reported in 1984.[1] In this report, excysted (free) sporozoites were inoculated onto human fetal lung cells and infection was detected by Nomarski interference contrast microscopy. Sexual stages (macrogametes and microgametocytes) were present by 48 hours and new oocysts were detected within 72 hours after inoculation. Following incubation in 2.5% potassium dichromate to kill all developmental stages, these cell culture-derived oocysts were demonstrated to be infectious for suckling mice. Complete development has also been reported in Caco-2,[2] RL95-2,[3] BFTE,[4] THP-1[5] and MDBK cells.[6] Oocysts were also observed in sporozoite-infected monolayers of HCT-8 and HT-29 cells.[7] However, no cell culture-based propagation procedure has been developed for *C. parvum* because the number of oocysts produced is low and always less than the number of oocysts in the original inoculum. The failure of *C. parvum* to propagate in cell culture is not yet understood but may be related to deficiencies in the culture medium or problems with the host cells resulting in the inability of the parasite to efficiently progress beyond the sexual stage of the life cycle.

There is currently no concensus on which cell line is the most appropriate for studying in-vitro development of *C. parvum*. A number of investigators have compared different cell lines to determine which supports the greatest infection. Upton *et al.*[10] compared 11 cell lines and reported that HCT-8 cells produced approximately twice as many intracellular life cycle stages compared to MDBK, MDCK, or Caco-2 cells.

Table 1   *Cell lines that support the in-vitro growth of C. parvum*

Cell line	Origin of cells	Reference
-	Mouse peritoneal macrophages	8
A-549	Human lung carcinoma	9
BALB/3T3	BALB/c mouse embryo	10
BFTE	Bovine fallopian tube epithelium	11
BS-C-1	African green monkey kidney	12
BT-549	Human breast infiltrating duct carcinoma	10
Caco-2	Human colonic adenocarcinoma	2
CHO	Chinese hamster ovary	13
H69	Human bile duct epithelium	14
HCT-8	Human ileocecal adenocarcinoma	10
HFL	Human fetal lung	1
Hs-700T	Human pelvic adenocarcinmoa	10
HT-1080	Human fibrosarcoma	10
HT-29	Human colonic adenocarcinoma	15
LS-174T	Human colonic adenocarcinmoa	10
MDBK	Bovine kidney	16
MDCK	Canine kidney	17
RL95-2	Human endometrial carcinoma	3
T84	Human colonic carcinoma	18
THP-1	Non-adherent human monocytes	5

However, other investigators reported that there was no difference in the amount of infection in either Caco-2, HCT-8 or HT29 cell lines.[7] Also, no difference was observed between HCT-8 and MDCK cells.[21] Selection of the most appropriate cells is not governed solely by the number of life cycle stages which develop in a particular cell line. For example, if complete development of the parasite is required with the production of de novo oocysts, then only cell lines such as Caco-2, RL95-2, or BFTE would be suitable, as discussed above. The method of infectivity detection may also influence the choice of cell line. It has been reported that MDBK cells are superior to HCT-8 when detecting infection by immunofluorescence microscopy (IFA) due to the high background fluorescence obtained with HCT-8 cells.[22]

Representation of the human intestine is also a factor in selecting cells. If an in-vitro infectivity assay is being used to predict the potential of waterborne oocysts to cause infection in humans, then the cells used should be a good model of the human intestine. For example, Caco-2 cells spontaneously differentiate once they reach confluency forming brush border microvilli and produce higher levels of brush border enzymes (alkaline phosphatase, sucrase-isomaltase and amino-peptidase) compared to non-differentiated cells.[23] This functional differentiation may make Caco-2 cells a better model of the human intestine than cell lines which do not possess these properties.

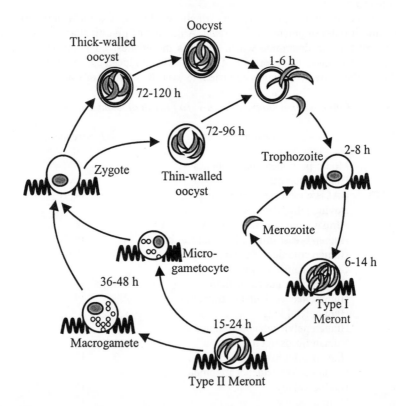

Figure 1 *C. parvum life cycle indicating post-infection periods (hours) at which different stages are detected in cell culture (based on published reports[2,3,5,15,19]). Life cycle diagram is adapted from Fayer et al.[20]*

HT29 cells can be induced to differentiate under glucose starvation conditions and differentiated HT29.74 cells (a clone of HT29) supported a five-fold increase in the development of *C. parvum* developmental stages compared to undifferentiated cells.[15]

Different cell lines require different media formulations for optimum growth. However, optimum cell growth conditions may not be conducive to maximum parasite development. Standard cell culture media, such as RPMI 1640 for HCT-8 cells or DMEM for Caco-2 cells, contain all of the essential nutrients for cell growth and are typically supplemented with fetal bovine serum (FBS) at concentrations ranging from 5% to 20%. Some investigators include antibiotics in the growth medium (100 U/ml penicillin, 0.1 mg/ml streptomycin, 0.1 mg/ml kanamycin, 0.25 µg/ml amphotericin) to suppress growth of contaminants while others believe it is not necessary. In one investigation a variety of media components and additives were compared and an optimized medium formulation was developed which supported a 10-fold increase in the number of developmental stages of *C. parvum* compared to standard media.[24] In addition to an RPMI 1640 base this parasite growth medium contained 10% FBS, 15 mM HEPES, 4 mM glutamine, 50 mM glucose, 35 µg/ml ascorbic acid, 1 µg/ml folic acid, 0.1 U/ml insulin, 4 µg/ml 4-aminobenzoic acid, 2 µg/ml calcium panto-thenate and antibiotics. There are conflicting reports on the optimum FBS concentration required once cells have been infected to allow maximum parasite development and it has been suggested that FBS may contain components that inhibit parasite growth.[25] The number of life cycle stages developing in MDCK cells grown in serum-free Ultraculture media (BioWhittaker) was approximately two- to three-fold higher when compared with cells grown in DMEM supplemented with 10% FBS.[25]

Following inoculation with oocysts, incubation times for development of infection typically range from 24 to 72 h. The longer incubation times should permit development of the sexual stages and new oocysts (Figure 1). For the purposes of this review an infectious foci is defined as a cluster of life cycle stages, presumably derived from a single invasive sporozoite. Studies using in-situ hybridization with a *C. parvum*-specific oligonucleotide probe to detect infection demonstrated that infectious foci contained an average of 4.2 ± 2.2 (n=106) stages after incubation for 24 h.[26] After 48 h incubation, the number of stages per foci had increased to an average of 25.7 ± 8.9 (n=190), a six-fold increase compared to 24 h. An investigation using IFA as the detection method reported 17-fold to 27-fold increases in the total number of stages after 48 h incubation compared to 24 h.[19]

3 ASSAY FORMATS AND DETECTION METHODS

A variety of infectivity assays have been developed for *C. parvum* using various cell lines, assay formats, and detection methods (Table 2). Any of these assays could be adapted to a variety of applications. The ELISA and chemiluminescence immunoassays in 96-well formats have been particularly useful for large scale screening of potential anticryptosporidial agents.[21,27] Woods *et al.*[27] screened the *C. parvum* inhibitory activity of 101 agents using HCT-8 cells and an ELISA detection method and generated dose-response curves for the most active compounds. The microscopic detection methods provide readily enumerated infectivity results because the actual life stages are visualized. However, such procedures can be time consuming. Molecular-based infection detection methods utilizing PCR to amplify either DNA or mRNA have also been developed.[12,28,29] These techniques are highly specific and sensitive and, as with the

Table 2 *Cell culture-based infectivity assays for C. parvum*

Cell line	Assay format	Detection method	Reference
RL95-2	Thermanox coverslips	Giemsa staining	3
MDBK	Glass coverslips	Interference microscopy	10
MDCK	96-well plates	Chemiluminescence immunoassay	21
MDCK	Chamber slides	Immunofluorescence	25
BFTE	Glass coverslips	Microscopy	11
Caco-2	Permeable membranes	Transmonolayer resistance	30
Caco-2	Chamber slides	RT-PCR on messenger RNA	28
HCT-8	96-well plates	ELISA	27
HCT-8	24-well plates	RT-PCR on messenger RNA	26
HCT-8	Chamber slides	Immunofluorescence	19,31
HCT-8	Chamber slides	In-situ hybridization	26
HCT-8	96-well plates	PCR on DNA	29
BS-C-1	24-well plates	PCR on DNA	12

enzyme-linked methods described above, can be used to screen a large number of samples simultaneously. An assay using RT-PCR to amplify *C. parvum*-specific mRNA from a region of the heat shock protein gene (*hsp*70) was used to detect infection in Caco-2 and HCT-8 cells with as few as 10 oocysts.[26,28] Messenger RNA was selected as the target for detection since it is actively produced by the parasite as it develops in cell culture, but is rapidly degraded if oocysts are incapable of initiating infection. Thus, the risk of false-positive detection of dead or non-infectious oocysts is greatly reduced. PCR-based methods which detect DNA may detect dead oocysts which are simply located on the cell monolayer but not actively infecting. DNA is far more stable than mRNA and may persist for extended periods inside dead oocysts. Although PCR and RT-PCR methods can be used for relative quantitation (which is useful for inactivation studies), it is difficult to generate the absolute quantitative data provided by microscopic methods. Consequently, an in-situ hybridization assay (ISH) which provides the specificity of molecular-based methods with the easy enumeration of microscopic methods was developed.[28] Preliminary investigations of dose response curves generated by ISH and RT-PCR demonstrated an almost perfect correlation and both methods correlated strongly with animal infectivity studies (P. Rochelle, unpublished).

The primary interest of the water industry in cell culture-based infectivity assays is to determine whether oocysts detected in environmental water samples are infectious and also to measure the efficacy of disinfection strategies. RT-PCR detection of infection on HCT-8 cells was used to demonstrate that oocysts recovered from environmental water samples by immunomagnetic separation and by USEPA Method 1622[32] retained their infectivity.[33,34] Also, using HCT-8 cells and PCR detection, 4.9% of raw water samples and 7.4% of filter backwash samples were found to contain infectious *C. parvum*.[29] The sensitivity of the latter assay was reported to be less than five infectious oocysts.[29] A sensitivity of a single infectious oocyst was reported for an assay using IFA to detect infection in HCT-8 cells.[19] However, standardization is necessary to ensure these reported sensitivities are obtained on a routine basis. *C. parvum* oocysts demonstrate considerable variation in infectivity, even with a single isolate prepared at different times. It has been established by many researchers that oocysts loose infectivity as they age and greater than 10-fold differences in cell culture infectivity have been reported for different lots of the IOWA isolate of the same age.[31] In a separate series of studies, also using the IOWA isolate and HCT-8 cell monolayers, infection was detected by RT-PCR in 41% of instances (n=17) when fewer than 25 oocysts were inoculated onto the cells.[26]

C. parvum oocysts are relatively resistant to standard disinfectants such as chlorine at commonly used concentrations. Consequently, alternative inactivation methods are required and cell culture-based infectivity assays have the potential for screening a wide range of inactivation agents under varying conditions. Recent work using RT-PCR detection of infection in HCT-8 cells has demonstrated that that pulsed-UV doses of 16 mJ/cm^2 provided >3 log_{10} inactivation of IOWA isolate oocysts.[35] More than 4 log_{10} inactivation was observed in HCT-8 cells (infection detected by IFA) for oocysts exposed to broad spectrum pulsed-white light.[31]

4 LIMITATIONS AND FUTURE DEVELOPMENTS

Although cell culture methods are been used in an increasing number of laboratories to study *C. parvum* infection, there is still a need for standardization of procedures and increasing the rate of infection. Higher rates of infection are achieved in cell culture if oocysts are pre-treated prior to inoculation onto cells. Such treatments include incubation in 10% bleach, acidification, or excystation in bile salts. However, these procedures may be detrimental to environmentally stressed oocysts and could reduce infectivity. Also, they could have synergistic effects when combined with disinfectants if oocysts are being used for inactivation studies. The final step in USEPA Method 1622 for recovery of oocysts from water involves disassociation of oocysts from the paramagnetic beads in 0.1 M HCl and it has been suggested that this procedure reduces the infectivity of oocysts in cell culture.[29] Incubation in bile salts prior to inoculation onto cells most closely mimics the conditions oocysts are exposed to in the human intestine, but there is currently no information concerning the interaction of bile salts and potential disinfectants. Therefore, more studies need to be conducted on the methods used to treat oocysts prior to infection.

There is increasing evidence that infection of cells by *C. parvum* induces apoptotic cell death in the host monolayers.[14,18,30] Following infection, host cell death was indicated by changes in cellular morphology, release of lactate dehydrogenase and decreases in transmonolayer electrical resistance within 24 to 48 hours of initiation of infection in Caco-2 and T-84 cells.[18,30] All authors reported that the extent of apoptosis was oocyst dose and time dependent. In H69 cells, apoptosis was detected with the DNA binding dye DAPI which demonstrates the characteristic nuclear changes associated with programmed cell death. Cytopathic effects were observed 48 h after infection and by 96 h 30% of the host cells were apoptotic.[14] Widespread cell death in infected monolayers may result in sloughing of cells from growth chambers, the elimination of *C. parvum* receptive cells from the monolayer, and may be partly responsible for the lack of a continuous auto-reinfection cycle in cell culture. Therefore, it is possible that inhibition of apoptosis in infected cells may extend the period that infection receptive cells are available and thus allow higher rates of infection. Ongoing investigations with apoptosis inhibitors in *C. parvum* cell culture will test these hypotheses (G. Widmer, pers. comm.).

5 CONCLUSIONS

Cell culture-based in-vitro infectivity of *C. parvum* is a useful tool which can provide valuable information about oocysts in the environment. It is a practical and relatively straightforward technique which can be used to measure the efficacy of disinfectants, the infectivity of waterborne oocysts, and the persistence of oocysts in environmental waters. It is a widely applicable technique which supports the growth of the two currently accepted genotypes of *C. parvum*. However, before the technique can be adopted by the water industry further research needs to be conducted comparing cell culture with the so-called "gold standard" of animal models. In addition, cell culture-based methods should be compared with human volunteers to determine their true sensitivity and predictive capacity. Also, cell culture methods, cell lines, and growth conditions need to be standardized and oocyst pre-treatments prior to infection need to be evaluated and standardized, particularly if cell culture is going to be used to study oocyst inactivation.

Acknowledgments

Parts of the work described in this paper were supported by a grant from the United States Environmental Protection Agency (award number R825146-01-0).

References

1. W. L. Current and T. B. Haynes, *Science*, 1984, **224** , 603
2. M. Buraud, E. Forget, L. Favennec, J. Bizet, J. Gobert and A. Deluol, *Infect. Immun.* 1991, **59**, 4610.
3. K. R. Rasmussen, N. C. Larsen and M C. Healey, *Infect. Immun.*, 1993, **61**, 1482.
4. S. Yang, M. C. Healey, C. Du, and J. Zhang, *Infect. Immun.*, 1996, **64**, 349.
5. P. Lawton, M. Naciri, R. Mancassola and A. Petavy, *J. Microbiol. Meth.*, 1996, **27**, 165.
6. I. Villacorte, D. de Graaf, G. Charleir, and J. E. Peeters, *FEMS Microbiol. Letts.*, 1996, **142**, 129.
7. C. Maillot, L. Favennec, A. Francois, P. Ducrotte and P. Brasseur, *J. Euk. Microbiol.*, 1997, **44**, 582
8. F. Martinez, C. Mascaro, M. J. Rosales, J. Diaz, J. Cifuentes and A. Osuna, *Vet. Parasit.*, 1992, **42**, 27.
9. A. Giacometti, O. Cirioni and G. Scalise, *Am. J. Trop. Med. Hyg.*, 1996, **38**, 399.
10. S. J. Upton, M. Tilley and D. B. Brillhart, *FEMS Microbiol. Letts.*, 1994, **118**, 233.
11. J. R. Forney, S. Wang, C. Du and M. C. Healey, *J. Parasitol.*, 1996, **82**, 638.
12. M. Q. Deng and D. O. Cliver, *J. Parasitol.*, 1998, **84**, 8.
13. K. S. Morrow and M. Dao, *Am. Soc. Microbiol. 99th General Meeting*, 1999, Abstract D/B-116.
14. X. M. Chen, S. A. Levine, P. Tietz, E. Krueger, M. A. McNiven, D. M. Jefferson, M. Mahle and N. F. LaRusso, *Hepatology*, 1998, **28**, 906.
15. T. P. Flanigan, T. Aji, R. Marshall, R. Soave, M. Aikawa, and C. Kaetzel, *Infect. Immun.*, 1991, **59**, 234.

16. S. J. Upton, M. Tilley, R. R. Mitschler and B. S. Oppert, *J. Clin. Microbiol.*, 1991, **29**,1062.
17. J. Gut, C. Petersen, R. Nelson and J. Leech, *J. Protozool.*, 1991, **38**, 72S.
18. R. B. Adams, R. L. Guerrant, S. Zu, G. Fang and J. K. Roche, *J. Infect. Dis.*, 1994, **169**, 170.
19. T. R. Slifko, D. E. Friedman J. B. Rose and W. Jakubowski, *Appl. Environ. Microbiol.*, 1997, **63**, 3669.
20. R. Fayer, C. A. Speer and J. P. Dubey, in *Cryptosporidiosis of Man and Animals*, R. Fayer, C. A. Speer and J. P. Dubey, CRC Press, Boca Raton, 1990.
21. X. You, M. J. Arrowood, M. Lejkowski, L. Xie, R. F. Schinazi and J. R. Mead, *FEMS Microbiol. Lett.*, 1996, **136**, 251.
22. S. Tzipori, *Adv. Parasit.*, 1998, **40**, 187.
23. M. Pinto, S. Robine-Leon, M. -D. Appay, M. Kedinger, N. Triadou, E. Dussaulz, B. Lacroix, P. Simon-Assmann, K. Haffen, J. Fogh and A. Zweibaum, *Biol. Cell.* 1983, **47**, 323.
24. S. J. Upton, M. Tilley and D. B. Brillhart, *J. Clin. Microbiol.,* 1995, **33**, 371.
25. M. J. Arrowood, L. Xie and M. R. Hurd, *J. Euk. Microbiol.*, 1994, **41**, 23S
26. P. A. Rochelle, R. De Leon, A. M. Johnson, D. M. Ferguson, H. Baribeau and M. H. Stewart, *Proc. AWWA Water Quality Technology Conference*, 1999, Amer. Water Works Assoc., Denver.
27. K. M. Woods, M. V. Nesterenko and S. J. Upton, *Annals Trop. Med. Parasit.*, 1996, **90**, 603.
28. P. A. Rochelle, D. M. Ferguson, T. J. Handojo, R. De Leon and M. H. Stewart, *Appl. Environ. Microbiol.*,1997, **63**, 2029.
29. G. D. Di Giovanni, F. H. Hashemi, N. J. Shaw, F. A. Abrams, M. LeChevallier and M. Abbaszadegan, *Appl. Environ. Microbiol.*, 1999, **65**, 3427.
30. J. K. Griffiths, R. Moore, S. Dooley, G. T. Keusch and S. Tzipori, *Infect. Immun.*, 1994, **62**, 4506.
31. T. R. Slifko, D. E. Huffman and J. B. Rose, *Appl. Environ. Microbiol.*, 1999, **65**, 3936.
32. United States Environmental Protection Agency, *Method 1622: Cryptosporidium in Water by Filtration/IMS/FA*, 1999, EPA 821-R-99-001, Office of Water, Washington.
33. P. A. Rochelle, R. De Leon, A. M. Johnson, M. H. Stewart and R. L. Wolfe, *Appl. Environ. Microbiol.*, 1999, **65**, 841.
34. P. A. Rochelle, R. De Leon, M. H. Stewart and R. L. Wolfe, *Proc. AWWA Water Quality Technology Conference*, 1998, Amer. Water Works Assoc., Denver.
35. A. Mofidi, H. Baribeau and J. F. Green, Proc. *AWWA Water Quality Technology Conference*, 1999, Amer. Water Works Assoc., Denver.

APPLICATIONS OF MALDI-TOF MASS SPECTROMETRY IN THE ANALYSIS OF *CRYPTOSPORIDIUM*

K. Hall[1], M.A. Claydon[2], D.J. Evason,[2] M. S. Smith[3] and J. Watkins[4]

[1]Hall Analytical Ltd.   [2]Bio-Analytical Research Centre   [3]Drinking Water Inspectorate
Floats Road   Manchester Metropolitan Univ   Ashdown House
Manchester M23 9YJ Manchester M1 5DG   123, Victoria Street
                                                                         London SW1E 6DE

[4]CREH Analytical Limited, Horsforth
Leeds, LS18 4RS

1   INTRODUCTION

Cryptosporidium parvum is an acknowledged pathogen associated with waterborne disease. It requires reliable methods of detection, quantification, identification, and viability to ensure appropriate water treatment strategies may be devised. In this study, we have investigated the use of MALDI-TOF mass spectrometry as a method for the identification of the *Cryptosporidium* oocyst. The approach taken was based on existing MALDI-TOF methods for the identification of microorganisms.[1-4] Such methods provide a mass spectrum of molecules desorbed from the cell wall giving a phenotypic fingerprint of the organism.

1.1 Methods

1.1.1 Collection and purification of oocysts. Two suspensions of the Moredun cervine isolate were purchased as cleaned viable suspensions in phosphate buffered saline (PBS). The batches tested were C1/98 and C3/98. The suspensions were cleaned by sorting on a flow cytometer (Becton Dickinson, FACS Vantage) using forward and side scatter parameters on an unstained suspension. The sorted material was collected in eppendorf tubes, centrifuged at 5,000 rpm for 2 minutes and resuspended in deionised water for analysis. A faecal sample of genotype 1 *C. parvum* was obtained from the Public Health Laboratory at Swansea. Oocysts were cleaned from the faecal material using an identical technique to that used by Moredun Scientific (Wright, pers. com.). This suspension was also further cleaned using flow cytometry.

1.1.2 MALDI-TOF analysis of oocysts. Two samples designated C1/98 and C2/98 were analysed together with the third sample of *C. parvum* genotype 1. The Moredon oocysts were obtained at a concentration 10^3 oocysts / ml and required further concentration before analysis. The genotype 1 sample had a concentration of 10^8

oocysts / ml and required no additional concentration prior to analysis. Concentration of the samples was achieved by centrifugation in a bench centrifuge for at 1,500 x g for 20 minutes in order to pellet the oocysts followed by centrifugation in eppendorf tubes as above.   The volume of the diluent was then reduced to 10μl by pipette and the pellet re-suspended.

The analysis began with a preliminary study to select a suitable matrix to assist in the laser desorption /ionisation of the samples. Two matrices were compared; α-cyano-4-hydroxycinnamic acid (α-CHCA) and 5-chloro-2-mercaptobenzothiazol (CMBT). One μl of each sample was pipetted on to a sample target in duplicate. The samples were allowed to air dry before the addition of 1 μl of saturated matrix (dissolved in water, methanol and acetonitrile 1:1:1) was added to each sample on the slide and also allowed to air dry before analysis. The samples were analysed on a Kompact MALDI 2 linear, time-of-flight mass spectrometer (Kratos Analytical, Manchester, UK.) and the analysis was repeated on a ToF Spec 2E (Micro Mass, Manchester U.K.) in linear mode. Both instruments use a nitrogen laser giving a 337nm output of 3ns pulse width. The

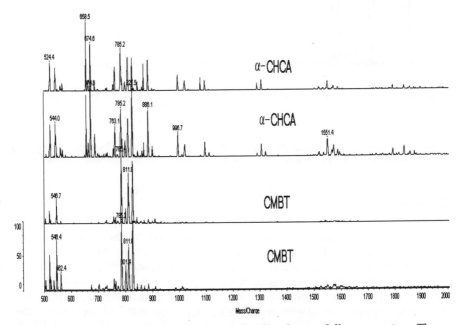

Figure 1 *Spectra of the Moredon samples analysed in the two different matrices. The α–CHCA matrix gives the widest range of information most reproducibly*

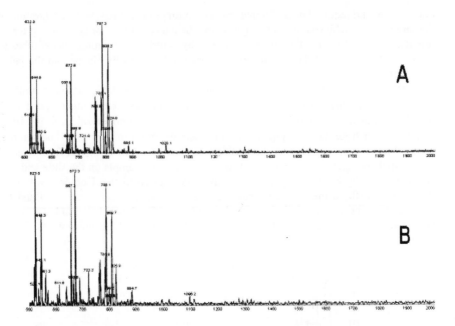

Figure 2 *The Moredon spectra obtained from (a)the Kompact 2 instrument and (b) the Tof Spec 2E instrument*

laser fluence was set just above the threshold for ion production for each instrument. The instruments were operated in the positive ion mode at an acceleration voltage of +20kV.

1.2 Results

1.2.1 The MALDI-TOF analysis of oocysts. The results of the matrix study are illustrated in Figure 1. This shows the duplicate spectra of the Moredon oocysts, obtained from both matrices. Although good spectra are obtained with either matrix, the data obtained using α-CHCA gave more detail and proved to be more reproducible than that obtained with CMBT. This tendency was also found in the other samples analysed. The α-CHCA matrix was adopted for the remainder of the study. Figure 2 shows the spectra of the same sample obtained from both instruments. It may be seen that both spectra are essentially the same. Figure 3 shows the spectra of the two Moredon oocysts together with the genotype 1 oocysts (m/z range 500-2000). The spectra from the Moredun oocysts are indistinguishable from each other but the spectrum of the genotype 1 oocysts displays a number of unique markers.

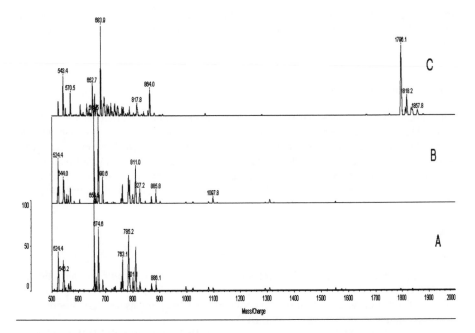

Figure 3 *The spectra of (a) Moredon Feb, (b) Moredon May and (c) Genotype 1*

1.3 Discussion

In this study approximately 10^5 oocysts were applied to each sample spot. One hundred spectra were gathered from the laser scan across the sample. These were summed to give an averaged spectrum representative of the population of oocysts. The spectra of the Moredon samples obtained in this way, appear identical. As the two samples were obtained from the same source at different times, it is highly likely that they are the same organism. It may be expected that their spectra are similar. The study was carried out on two different instruments to determine whether the method was instrument dependant. The results obtained suggest that the method can be applied to different instruments and comparable results obtained.

When the spectra of all three samples are compared it can be seen that the Moredon samples can be easily distinguished from the genotype 1 sample. The Moredon samples are both genotype 2. This implies that *Cryptosporidium* species may be identified and sub-divided into genotypes using the MALDI-TOF method. The preparation and analysis time for each sample is a few minutes. This makes the technique a rapid method, suitable for screening a large number of samples. Although the number of samples used in this study is small and a wider panel of samples is required to confirm the results presented here, the method shows potential as a means to identify *Cryptosporidium.*

1.4 Acknowledgements

The authors wish to thank the Drinking Water Inspectorate (DWI), Hall Laboratories and CREH *Analytical* limited for permission to publish this paper. We also wish to thank Alcontrol Laboratories for the use of their FACS Vantage flow cytometer.This work was undertaken as part of a research contract funded by the Department of the Environment, Transport and the Regions and managed by the DWI. The opinions expressed in this paper are those of the authors and not necessarily the companies that they represent or the DWI.

1.5 References

1 M.A. Claydon, S.N. Davey, V. Edwards-Jones & D.B. Gordon, *Nature Biotec.,* 1996, **14**, 1584.
2 R.D.Holland, J.G. Wilkes, F. Rafii, J.B. Sutherland, C.C. Persons and K.J. Voorhees, *Rapid Commun Mass Spectrom*, 1996, **10**, 1227.
3 V. Edwards-Jones, M.A. Claydon, D.J. Evason, J. Walker, A. Fox and D.B. Gordon, *J. Med. Micro.*, 1999 (in press).
4 N.H. Shah, S.E.Gharbia, C. Keys, K. Ralphson, F. Trundel, I. Brookhouse, and M.A. Claydon, *Microbial Ecology in Health and Disease*, 1999 (in press).

SOME OBSERVATIONS ON FACTORS WHICH AFFECT RECOVERY EFFICIENCY IN *CRYPTOSPORIDIUM* ANALYSIS

A.P.Walker
North West Water Plc
Lingley Mere Laboratory
Lingley Green Avenue
Great Sankey
Warrington   WA5 3QT

Abstract

A wide range of reasons for poor recovery efficiency of *Cryptosporidium* oocysts from waters during concentration, extraction, purification, identification and counting have been identified by many workers in the field. Observations and data are presented here which look at some of the fine detail of technique which may significantly affect how many oocysts are either lost or fail to be detected at individual stages of analysis. In particular, the preparation, fixing, staining, rinsing, aspirating, mounting and examination of microscope slides has come under scrutiny in an attempt to close this particular loophole in the analytical process.

1 INTRODUCTION

Loss of Cryptosporidium oocysts during analysis affects virtually every technique that has been developed. Much has been made of the potential for oocyst loss during the many sequential stages of the various mainstream methods for isolation, purification identification and counting of Cryptosporidium (1,2,3). Some typical recoveries are given in Table 1.

Table 1 *Typical recoveries of oocysts at individual stages*

Filtration and Extraction (overall)	2% - 96% (Ref. 2)
Filtration and Extraction (Flatbed membrane)	38% - 56% (NWW)
Centrifugation	32% - 92% (NWW)
Gradient/Flotation	1% - 60% (Ref. 2)
Gradient/Flotation	Mean 56% (NWW)
Cytometry	Up to 100% (NWW)
Immunomagnetic Separation	Up to 100% (NWW)
Slide Staining/Mounting	30% - 100% (NWW)

Losses at the final stage, on the slide, is arguably one of the most unpredictable causes of unexplained sporadic losses during Cryptosporidium recovery. Most analysts who have carried out replicate spots on slides will have experienced unexpectedly low wells e.g. 75, 72, *23*, 84, 66, 71 etc.. It has been suspected for some time that significant losses could be occurring during fixing, staining, stain removal/rinsing and mounting stages. In an attempt to reduce this risk, aspiration has been adopted as the method of choice for stain and rinse liquid removal from slide wells in both the Blue Book (2) and the Standard Operating Protocols (SOPs) for the new Cryptosporidium Regulations (4).

Loss of oocysts from the slide has the potential to seriously affect the results of all analytical and investigational work that relies on final mounting of oocysts on slides. In effect, our most carefully designed experimental work has the potential to give totally erroneous results if we do not pay sufficient attention to detail at this crucial final stage. This includes the SOP regulatory methods, and in particular, control slides. It could also affect any work carried out to investigate individual stages of the analytical process, for example the stages mentioned in Table 1. Few papers appear to have given a great deal of consideration to the possibility of losses from the slide, (apart from early work by Musial et al (5) relating to the use of egg albumin and poly–L lysine slide coatings as sticking agents – and these appear not to have retained favour).

It therefore follows that a serious investigation of factors affecting losses from slides could help to increase the reliability of overall analytical and investigational work. This paper has therefore concentrated on studying the causes of losses from slides.

2 MATERIALS AND METHODS

All oocyst preparations used in this work were obtained as heat inactivated preparations from Moredun Scientific Ltd., Pentlands Science Park, Bush Loan, Penicuik, Midlothian, EH26 0PZ. The working preparations were coded with the Moredun concentrate batch code (e.g.C1/99) and given an extra identifying digit (e.g. C1/99/1).

Working suspensions of oocysts were prepared containing approx. 100 oocysts were prepared by adding 20 microlitres of a 10^7per 2ml Moredun preparation to 10ml of Phosphate Buffered Saline (PBS) in a 50ml Becton-Dickinson "Falcon" centrifuge tube, and vortex mixing thoroughly.

Mean counts of working suspensions were checked by spotting onto ImmunoCell 3-well Teflon coated diagnostic slides type 61.100.64 (Obtained from Merck Ltd., Merck House, Poole, Dorset, BH15 1TD).

Slides used for other work as indicated were either Dynal Spot-on slides , or Genera Quanti-Max well slides. Dynal slides from Dynal Ltd., 11, Bassendale Road, Croft Business Park, Bromborough, Wirral, CH62 3QL. Genera slides from Genera Technologies Ltd., Lynx Business Park, Newmarket, Cambs., CB8 7NY.

All oocyst slides were MAb stained either with TCS/Cell-Labs Crypto/Giardia Z1RR2 stain, or with TCS/Cell-Labs Crypto-Cel Z1RR1 (Crypto only) stain. Staining was carried out on slides for 60-90 minutes at 37°C with 50 microlitres of stain per well and aspirated as per the SOPs.

Staining in suspension for the flow cytometer was carried out at 37°C for 30 minutes.

DAPI staining was carried out with a freshly prepared DAPI solution in PBS prepared as per the SOPs.

Rinsing after staining/DAPI was carried out by adding 2 drops of Reverse Osmosis (RO) water and aspirating as per the SOPs.

Mounting of slides was carried out with TCS IFA mounting fluid Z1MM10, one drop per well.

Flow Cytometry was carried out using a Coulter Elite flow cytometer to sort either unstained oocysts, with selection criteria based on side and forward scatter, or stained oocysts with additional selection based on fluorescence.

3   EXPERIMENTAL WORK

The following questions were considered:-

- Spotting – is the slide wettable or hydrophobic? Does this affect ability to stick?
- Fixing – does it work? Is methanol or acetone better than no fixing at all?
- Staining – Is the stain effective?
- Removal of stain
    - does gentle aspiration really prevent loss?
    - what if the vacuum is too strong?
    - Does DAPI and the subsequent rinse droplet increase the loss?
    - Is it affected by the type of aspirator tip used?
- Mounting
    - Shear forces under cover slip during mounting (SOPs say no pressure)
    - What happens if high pressure *is* used?

The experimental work was carried out in two stages:-
- Earlier work was carried out on hydrophilic slides only using 10+ replicates due to statistical variation.
- The later work on hydrophobic and hydrophilic slides was carried out using flow cytometric sorting, reducing the need for numerous replicates.

4 RESULTS

4.1 Effect of Fixative on Hydrophilic Slides

The comparative effects of methanol, acetone and no fixative (i.e. air-drying only) were examined using hydrophilic slides. A small amount of sediment from a upland raw water was included in a repeat of the experiment to see whether fixative performance was affected by presence of typical detritus. The results are given in Table 2.

No significant differences are evident between the two fixatives and no fixative, either with or without sediment. Methanol was therefore selected for all subsequent work in line with the SOPs, although there would appear to be no advantage in using a fixative with hydrophilic slides.

Table 2 *The effect of fixative on hydrophilic slides*

	No Fixative	**Methanol**	**Acetone**
Without Sediment			
Mean	**169.3**	**152.5**	**156.5**
Standard Deviation	10.6	26.8	21.1
No. Replicates	11	11	11
With Sediment			
Mean	**123.1**	**133.3**	**129.9**
Standard Deviation	10.0	14.6	22.1
No. Replicates	11	11	11

4.2 Effect of Aspiration v Rinsing

A smaller experiment with fewer replicates was done to establish whether aspiration was better than rinsing on hydrophilic slides.

Aspiration was carried out using a vacuum of <5cm Hg with a water pump vacuum source and a micropipette tip (Eppendorf 1000microlitre size) as the aspiration tip.

Rinsing was carried out using a wash bottle playing a jet of liquid onto the slide to one side of the well (not directly onto the well) and allowing a gentle stream of rinse liquid to flow across the well with the slide tilted at a slight angle.

Results are given in Table 3.

Table 3 *The effect of aspiration vs. rinsing*

	Rinsed		Aspirated	
	No Fixative	Fixed (Acetone)	No Fixative	Fixed (Acetone)
Mean	**11**	**10**	**38**	**32**
Standard Deviation	6.04	5.00	5.52	8.09
No. Replicates	3	3	3	3

The results here are very clear cut, with the counts on the aspirated slides being three times greater than the rinsed slides. Rinsing is clearly a non-starter, even on the relatively good adhesion exhibited by hydrophilic slides.

4.3 Effect of Aspiration at Low and High Vacuum

The SOPs state that a vacuum of <5cm Hg must be used. This experiment was designed to compare low vacuum with the other extreme of 30cm Hg which was achieved with a powerful water venturi vacuum pump. Micropipette tips (Eppendorf 1000microlitre size) were used as aspirator tips. Results are given in Table 4.

Table 4 *The effect of aspiration at high and low vacuum*

	Aspiration at <5cm Hg	**Aspiration at 30cm Hg**
Mean	**142**	**125**
Standard Deviation	18.21	24.49
No. Replicates	11	11

Although the differences in the means are not large, a t-test indicated that there was a significant difference between the two sets of data at the 5% level, suggesting that low vacuum aspiration caused less oocyst loss than high vacuum.

4.4 Preparation of Fixed Count Slides by Flow Cytometer

Slides were prepared by sorting 50 oocyst aliquots unstained onto Dynal and Genera slides. The sorted spot was deposited close to one edge of the well due to the central positioning of the single well on Dynal and Genera slides.

The slides were then methanol-fixed, stained with Crypto MAb batch RR163A, DAPI, then 1 drop of RO water to rinse, dried at 37°C and mounted without application of pressure. Aspiration was as before with a 1000microlitre tip. Care was taken to draw the liquid from the edge of the opposite side of the well to the sorted spot of oocysts, to minimise the risk of oocyst loss.

One clear difference was observed between the ImmunoCell 3-well slides and the Dynal and Genera slides as they were being prepared.. The glass within the ImmunoCell slide well was not hydrophobic and the sorted drops remained spread as they dried. In contrast, the hydrophobic nature of the Dynal and Genera well glass caused the sorted drop of liquid to shrink in area as it dried, taking the oocysts with it. The result was that oocysts ended up in a very small area with oocysts jammed together in ridges of salt. This appeared to prevent direct contact between many oocysts and the glass surface. It is therefore likely that this interfered with the effective fixing of the oocysts to the glass surface when the methanol was applied.

Microscopical examination was carried out before and after mounting to look for the effect of very gentle mounting on the oocyst positions. Results are given in Table 5.

Table 5 *Hydrophilic slides seeded by flow cytometer – losses and movement of oocysts during staining*

	Oocyst count after staining (a)	Oocyst count after mounting (b)	Oocysts in original positions (with halos) (c)	Moved oocysts (no halos) (d)	Halos without oocysts (e)	Oocysts Lost (f)
% of oocysts sorted	74	75	47	28	50	25
Range	56-96	60-94	22-58	6-82	34-82	0-44
No. of replicates	5	5	5	5	5	5

The data revealed that under the gentle mounting conditions applied, the oocysts did not move during mounting (compare columns a and b. The slight increase in apparent count is due to the image being clearer after mounting). However, many oocysts had moved away from the original spot left by the flow cytometer sort during the staining/aspiration process. They appeared to have become detached from the sorted spot which was offset to one side of the well, and drawn as a swathe across the well by the flow generated by aspiration from the opposite side of the well (see column d). Oocysts that had not moved had formed halos of tiny fluorescent spots where they had been stained *in situ* (column c) whereas oocysts that had moved were devoid of any halo and were outside the sorted spot (column d). The delineation of the original sorted spot was clearly visible under the microscope.

Oocysts that had moved from their original position left behind them an empty halo, most of which were clear enough to count (column e).

It was therefore possible to get a good estimate of the proportion of oocysts that had become detached during the staining process, and of those, how many had been lost completely by being aspirated off the slide (column f).

The mean loss of 25% (range from 0-44%) shows a serious potential for loss at this stage. The fact that up to 82% had moved from their original positions shows an even greater potential for loss if the aspiration is carried out more vigorously.

4.5 Hydrophobicity Comparison

In view of the apparent hydrophobicity shown by some slide types, examples of Genera, Dynal and ImmunoCell slides were compared. Two drops of flow cytometer sheath fluid were placed in a well of each slide type and the degree of spread compared. The drops spread to approx. 1mm from the edge of the well on both Dynal and Genera slides (no discernible difference). Tilting the slides caused the drop to move to the edge of the well on the downhill side but away from the edge on the uphill side.

However, the ImmunoCell slide, despite having larger (11mm) wells, was more hydrophilic, and two drops covered the entire well, with no tendency to draw away from the edge of the well when tilted.

4.6 Reducing Losses During Aspiration from Hydrophobic Slides using Fine Tips

The diameter of the fine hole in the end of an Eppendorf 1000microlitre aspiration tip was measured at 0.75mm, giving an orifice area of $0.44mm^2$. The flow rate generated by vacuum suction of less than 5cm Hg was observed to be fast, taking rather less than half a second to draw away 50 microlitres of stain. Cytometer-sorted slides each having 50 unstained oocysts were therefore stained following the same procedure, but using 100 microlitre tips instead. These had a fine hole diameter of 0.49mm (area $0.19mm^2$). This was found to noticeably slow the flow rate off the well. Counts were carried out to look for losses and are shown in Table 6. Counting was done unmounted.

Table 6 *Hydrophobic slides seeded with unstained oocysts by flow cytometer – effect of fine-tip aspiration on oocyst movement during staining*

Slide Number	Oocysts in original positions with halo in sorted area	Detached oocysts no halo outside sorted area	Total oocysts
1	46	0	46
2	47	0	47
3	47	0	47

This approach appears to have been effective, with counts much closer to the target 50. The apparent losses are down to 6-8%, and the losses could be due to a small proportion of particles other than Crypto being sorted by the unstained cytometry method. This view is supported by the observation that no oocysts had moved from their original positions.

4.7 Effect of Mounting Technique on Oocysts stained on Hydrophobic Slides

The slides used to generate the unmounted counts in Table 6 were then mounted in different ways. Slide 1 was mounted very heavily using warm mounting fluid. The cover slip was pressed down at one end by pressure of both index fingers whilst the other end was held up by the thumbs, then released suddenly. This action caused a very rapid shear flow across the well from the side where the sorted spot was, towards the opposite side.
Slide 2 was mounted very gently, by lowering vertically and allowing the mountant to spread slowly under the weight of the cover slip. Slide 3 was mounted at an angle with intermediate force. The results are given in Table 7. Clearly the heavy handling has led to movement of the oocysts, the heavier the handling, the greater the movement.

Table 7 *Hydrophobic slides seeded with unstained oocysts by flow cytometer – effect of mounting technique on movement of oocysts*

Slide Number	Oocysts in original positions with halo in sorted area	Detached oocysts no halo outside sorted area	Total oocysts
1 Heavy mount	35	11 *	46
3 Moderate mount	42	5	47
2 Gentle mount	47	0	47

* 6 out of 11 oocysts were off edge of well

4.8 Effect of Sorting Stained Oocysts onto Hydrophobic Slides

Sorting stained (rather than unstained) oocysts onto hydrophobic slides was found to overcome the hydrophobicity of the glass such that the sorted spot remained spread as it dried. This prevented the oocysts from becoming tightly packed together. It also prevented the oocysts from being lifted from the glass surface in ridges of salt as had happened previously with the unstained oocysts.

Clearly something in the liquid phase of the sorted drop was acting as a surface-active agent enabling the hydrophobic glass surface to be wetted. Since the oocysts for the cytometer are stained in suspension, each sorted droplet would have contained some stain in the liquid surrounding the oocyst. It is presumably some component of this stain that was responsible for the wetting effect.

A number of slides were cytometrically seeded with 50 stained oocysts in this manner and subjected to a range of fixing and mounting treatments. The slides were examined at several stages through the process (without mounting except at the end) to identify accurately where losses may be occurring. The oocysts were stained prior to flow cytometry and so could be counted without further staining.

Results are given in Table 8.

Table 8 *Hydrophobic slides seeded with stained oocysts by flow cytometer – effect of stain on adhesion of oocysts*

Slide Treatment	Count after Staining			Count after Fixing			Count after Mounting		
	Slide 1	Slide 2	Slide 3	Slide 1	Slide 2	Slide 3	Slide 1	Slide 2	Slide 3
Methanol Fix / Light Mount	50	50	50	50	49	50	50	50	50
Methanol Fix / Heavy Mount	50	50	49	50	50	49	50	50	49
Acetone Fix / Light Mount	50	50	50	50	50	50	50	50	50
Acetone Fix / Heavy Mount	49	49	50	49	50	50	50	50	50
Methanol Fix / Stain / DAPI / Light Mount	49	48	50	49	50	50	49	50	50
Methanol Fix / Stain / DAPI / Heavy Mount	50	50	48	50	50	48	50	50	48

Where mounting is described as "heavy", it describes a process of angled dropping of the cover slip followed by turning the slide over on paper towel and exerting full body-weight pressure to squeeze out excess mountant as rapidly as possible. This was designed to represent a "worst case scenario" for ham-fisted mounting.

Clearly the strength of adhesion is very strong when stain is present prior to the initial drying on the slide, even when the slide glass is hydrophobic. Even the heaviest mounting could not move these oocysts. Similarly, going through the full MAb stain / DAPI process on these slides had little or no effect on the oocyst counts.

There is clearly a crucial difference between the adhesion of the stained and unstained slides which appears to hinge round the presence of stain in the liquid drop enabling the hydrophobic glass to be wetted.

5 CONCLUSIONS

1. Oocysts can be lost at each stage of the analytical process.

2. Up to 70% of oocysts can be lost from slides during staining and mounting.

3. Rapid aspiration can cause loss if the oocysts are inadequately fixed to the glass – speed of liquid removal rather than vacuum strength *per se* appears to be critical.

4. Heavy mounting can cause loss if the oocysts are inadequately fixed.

5. Losses are greatest where the slide is hydrophobic.

6. The best attachment is achieved if the glass is effectively wetted by the oocyst suspension before drying.

References

1. Fricker C.R., *Why Can't We Detect Crypto?* Panel Discussion, Proceedings of 1997 International Symposium on Waterborne Cryptosporidium, AWWA, 385-402.

2. Methods for the Examination of Waters and Associated Materials - *Isolation and Identification of Cryptosporidium oocysts and Giardia cysts in Waters.*, Environment Agency, 1999.

3. Smith H.V., *The Status of UK Methods for the Detection of Cryptosporidium spp. Oocysts and Giardia spp cysts in water concentrates.* 1997, **35,** No.11/12, 369-376.

4. Standard Operating Protocols for the Monitoring of *Cryptosporidium* Oocysts in Water Supplies to satisfy Water Supply (Water Quality)(Amendment) Regulations 1999. Drinking Water Inspectorate.

5. Musial C.E., Arrowood M.J., Sterling C.R. and Gerba C.P. *Detection of Cryptosporidium in water by using polypropylene cartridge filters. Appl. and Environ. Microbiol.* 1987, **53,** 687-692.

DEVELOPMENT OF A NOVEL METHOD FOR THE CAPTURE, RECOVERY AND ANALYSIS OF *CRYPTOSPORIDIUM* OOCYSTS FROM HIGH VOLUME WATER SAMPLES

A.C.Parton, A.Parton, B.Brewin, K.Bergmann, E.Hewson, D.Sartory*

Genera Technologies Ltd., Lynx Business Park, Fordham Rd., Newmarket, Cambridgeshire, CB8 7NY
*Severn Trent Water, Quality and Environmental Services, Shrewsbury, UK

1 INTRODUCTION

The presence of *Cryptosporidium* oocysts in the water supply has resulted in a number of outbreaks of cryptosporidiosis in recent years, Swindon and Oxfordshire, UK[1] and Milwaukee, USA[2]. These outbreaks have highlighted the requirement for an efficient *cryptosporidium* oocyst monitoring system that is rapid, reliable and easy to use. With the introduction of government regulations in the UK pertaining to the presence of *Cryptosporidium* oocysts in drinking water, coming into effect early in the year 2000, the procedure for capture, recovery and enumeration has come under close scrutiny. Standard methods are poor with highly variable recoveries and lack of sensitivity. Until recently such monitoring has involved the trapping of oocysts using depth filters[3] or membrane filters[4,1]. Depth filters, whilst accommodating a high sample throughput, suffer with oocyst breakthrough and poor recovery. Membrane filters efficiently capture oocysts, but are very limited in the sample volumes that they can handle, with recoveries decreasing dramatically as sample volume and/or turbidity increase. Overall, the process of oocyst recapture from these filters for enumeration has proved to be labour intensive and time consuming[10] leading to highly variable recovery efficiencies[9]. The only alternatives, flocculation with calcium carbonate[11] or vortex flow and cross flow filtration[12] were found to be impractical for routine and outbreak monitoring purposes. A number of methods currently exist for the purification of oocysts from the sample debris with density gradient centrifugation, flow cytometry and immunomagnetic separation (IMS) being widely used. However low recovery efficiency, complexity of use and cost are all drawbacks to the use of these systems.

Enumeration of oocysts currently involves lengthy, tedious manual counting using a standard fluorescence microscope. Microscopic analysis of *cryptosporidium* oocysts requires a high degree of operator skill, judgement and accuracy to avoid errors associated with misidentification and scanning. The UK regulations have highlighted the need for increased traceability of counts, together with the ability to rapidly transfer data for external verification. In addition, many of the laboratories carrying out routine testing will be faced with large numbers of samples to evaluate, and as a consequence, microscope operators may be subject to fatigue.

The water industry is in need of a simple, efficient and reliable procedure for the capture, recovery and analysis of *Cryptosporidium* oocysts from water. Genera Technologies have developed a novel filter capture system, an IMS concentrate clean-up system and a semi-automated detection system, which when combined can provide a simple, low cost and efficient test for the detection of *Cryptosporidium* in water. The combined system provides reproducible results and is not dependent upon operator skill or analysis laboratory competence.

The capture system, Filta-Max™, consists of a compressed foam filter which incorporates 60 reticulated foam discs. The filter modules are very compact and fit into a housing approximately 10cm deep, thus making them ideal for use in the field. In this compressed state the foams act as a barrier filter, trapping *Cryptosporidium* oocysts. When the filters are allowed to expand the pores open and allow an elution solution to wash the foam efficiently. The Filta-Max™ is capable of filtering 2,000 l of finished water at flow rates of 3-4 l/min, far in excess of membrane alternatives. The IMS concentrate clean up recovery system, Puri-Max™ specifically captures the oocysts from the Filta-Max™ concentrate. This IMS system is unique in that the beads are immobilized onto a small capture phase, located in a magnetic field, over which the sample is circulated repeatedly, so increasing the chances of capture. The oocysts are captured onto the phase via antibody coated paramagnetic beads and the sample excess washed away. Oocyst recovery is by simple removal from the magnetic field. The QMS, Quanti-Max™ analysis system addresses many of the problems associated with manual microscopy by increasing the efficiency, accuracy and reliability of microscopical examination as well as reducing operator fatigue. *Cryptosporidium* oocyst enumeration is achieved using a PC-operated immunofluorescent scanning system (comprising an automated stage and focus drive) microscope (including fluorescence and DIC (Differential Interference Contrast) microscopy). Samples are scanned systematically field by field, with the user performing focus operations where necessary. During this initial scan, the user selects those features which are potential *Cryptosporidium* oocysts. The x, y and z coordinates for each feature are then stored. A review process of each stored feature, at high magnification, allows accurate classification of presumptive oocysts using FITC and DAPI stains and DIC microscopy. At each stage of the review, positive classification results in each image being stored on the PC harddrive, enabling an accurate physical map to be created for future reference. In addition full traceability of oocyst counts is provided with automatic storage and complete data logging of FITC, DAPI and DIC images. QMS is easy to use, with Wizard-based software guiding the user through every stage of the procedure. A program camera control unit with several pre-set shutter and light levels negates any requirement for complex shutter speed/light level determination by the operator.

2 FILTA-MAX™

2.1 Foam Filtration Units

The Filta-Max™ wash station consists of a foam filter module comprising 60 open cell reticulated foam discs (55 mm outer diameter and 15 mm inner diameter) sandwiched between two retaining plates and compressed by tightening a retaining

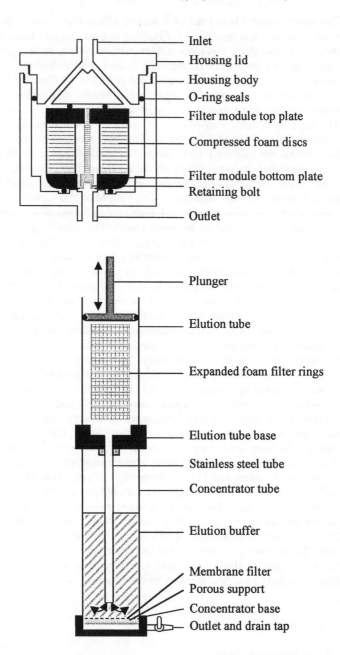

Figure 1 *Filta-Max^{TM}*

bolt. This module fits into a filter housing which has a screw top and seal (Figure 1). The filter housing incorporates a fitting which allows it to be simply attached to a water supply sampling point.   The water flows into the housing between the compressed foam discs and the housing wall, through the compressed foams (40 mm path length) into the centre space and out through the outlet port.

2.2 Elution Procedure

The Filta-Max™ wash station consists of an upper elution tube with plunger, and a lower concentrator tube, both of which screw into an elution tube base. The base fits into a wash station clamped to the bench (Figure 1). The lower concentrator tube has a 3 μm pore-size cellulose nitrate membrane filter fitted and a closed drain tap in its base.   Once the desired sampled volume has been filtered, the filter module is removed from the housing and screwed into the base of the plunger head. The plunger head is then lowered into the elution tube and locked into the lower position. Using an allen key, the retaining screw is removed. A 600 ml aliquot of a wash buffer (PBST) is added to the concentrator tube and the wash station assembled as shown in Figure 1.   The plunger head is then pumped 20 times.   During each compression/expansion cycle liquid is drawn into the foams as they expand and is expelled as they are compressed. After 20 cycles the plunger is locked in the lowest position to ensure the collection of the entire eluate in the concentrator tube. The concentrator tube is then removed and fitted with a suspended magnetic stirrer bar and placed on a magnetic stirrer. The stirrer is switched on and the eluate drained out through the membrane via the drain port until approximately 20 ml remains. Constant stirring ensures that the *Cryptosporidium* oocysts remain in suspension and are not filtered down onto the membrane. A hand held vacuum pump and drain reservoir aid this concentration process. The procedure is then repeated with a second 600 ml wash to which the first 20 ml concentrate is added. The final 20 ml concentrate is decanted into a 50 ml centrifuge. The membrane filter from the base of the concentrator tube is placed in a small sealable polythene bag with 5 ml extraction solution and the surface rubbed. The solution from this washing is added to the concentrate. The concentrate is then ready for Puri-Max™.

3 PURI-MAX™

3.1 Circulation

Paramagnetic beads coated with anti-*Cryptosporidium* antibody are added to the Filta-Max™ concentrate along with a blocking agent (2 % bovine serum albumen). The concentrate is then added to the Puri-Max™ reservoir located in the Puri-Max™ workstation (Figure 2). The tube and in-line filter are then put in place with the inlet and outlet tubes below the liquid level in the reservoir and the tube system fed through the pump head. The system is switched on and the solution allowed to circulate for 25 min at a pump speed of 300 rpm. The reticulated foam in the in-line filter enables the sample to be cleaned prior to the separation stage.

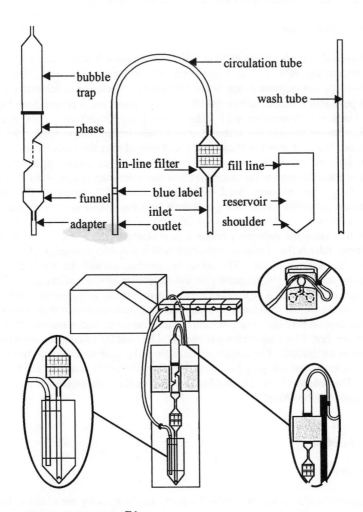

Figure 2 *Puri-Max*^{TM}

3.2 Separation

The pump is switched off and the magnetic phase is introduced just above the in-line filter and placed into the cradle within the magnet. The pump is then switched back on (reduced speed 100 rpm) and the system allowed to fill, once the system is running well the pump speed is increased to 300 rpm. Separation is performed at 300 rpm for 55 min.

3.3 Washing

The pump is switched off and the inlet tube transferred to a wash reservoir containing 500 ml wash buffer (PBST). The outlet tube is placed in a waste pot and the pump switched back on. Washing is performed until the wash reservoir is empty but is stopped before air enters the system. The inlet tube is removed and a 50 ml centrifuge tube placed below the phase in the magnet. The tube attachment above the phase is then disconnected and the solution drained into the centrifuge tube. Both the magnetic phase and the centrifuge tube are removed from the magnet. A small adapter tube is attached to a 10 ml syringe which is itself then attached to the phase and the beads are eluted from the magnetic phase by vigorous flushing.

3.4 Sample Clean Up and Enumeration

The entire eluate from each run is then filtered across a 13 mm, 2 μm pore size Isopore membrane filter (Millipore). Excess fluid is gently flushed out with air using a syringe, prior to staining with Crypt-o-Glo (Waterborne Inc.) FITC-monoclonal antibody. Enumeration of the *Cryptosporidium* oocysts is performed by standard immunofluorescence microscopy by counting the oocysts present on the Isopore membrane. Alternatively, the stained oocysts are removed from the membrane by rubbing the membrane surface in distilled water in a small plastic bag. The wash is transferred into an Eppendorf tube and centrifuged, the supernatant removed and the remaining sample volume transferred to a prewarmed 9 mm glass well slide (Genera Technologies) and the sample methanol fixed. Enumeration is achieved using a Nikon E600 fluorescence microscope with a 20x fluor objective, fitted with a Nikon fluorescien isothiocyanate (FITC) B-2A filter block, excitation 450-490 nm, emission 520 nm, for manual counting and the QMS™ for automated counting.

4 QUANTI-MAX™

The QMS™ system comprises, a Nikon E600 fluorescent microscope with Differential Interference Contrast (DIC) attachments together with a standard UV-A (DAPI) filter block and a Nikon B-2A (FITC) filter block (an excitation filter of 450-490nm and a barrier filter at 520nm). The optical arrangement comprises 1x, 20x lens and a 100x (oil immersion) lens. A JVC KY-F58, 3CCD, 750 lines resolution colour video camera (RGB output) is attached to the microscope using a 0.45x C-mount. Power is supplied to the camera via a JVC Camera Control unit (KY-F58) with lighting levels/shutter speeds, determined by programming the Digital Memory function of this unit.

Automation of the x, y and z stage movement has been achieved by attaching a Prior Scientific H101- automated stage with slide holder and a Prior Scientific fine focus motor to the Nikon E600 microscope. A PC (333 MHz Intel Pentium II processor with 64 MRAM, a 4.3 GByte hard disk drive) running a Windows 98 operating system controls the automated movement and the image capture functions. The PC has Autostage and Autofocus drive boards installed together with an ISDN or 56K modem interface. A Kensington Expert Mouse Rollerball is used to control the manual focus. A 17-inch high-resolution PC monitor is used to view the output from the 3CCD JVC video camera.

The image capture requirements and live image interaction have been custom designed by Genera Technologies in collaboration with Leica Microsystems Imaging Solutions with the resulting hardware and software installed on the QMS™ system.

The ability to view and record all live sequences seen on the QMS PC monitor has also been included in the system specifications. For this purpose a Matrox Image processor board with Matrox rainbow runner and CU-seeMe cards, has been installed. This allows for SVHS quality images to be viewed and recorded on a SVHS JVC 17-inch monitor and SVHS video recorder respectively. The inclusion of a modem into the specification allows data and images to be electronically transferred.

In order to illuminate the label portion of a slide, a Prior dual wand cold light source is included with the system specification.

The QMS™ system is easy to use. After switching on, the screen prompts are followed. The microscope stage is initialized and the slide label name scanned and recorded. The slide is then automatically scanned field by field, the user being required only to click on each presumptive *Cryptosporidium* oocyst. These images and their location are then stored by the PC, so that having completed the analysis, the review process can be performed. Each image is then reviewed this time under 100x oil immersion, and FITC, DAPI and D.I.C results recorded. A complete map of the scan area is then produced allowing complete analysis history of the stored data.

5 DATA

Table 1 shows that the combined Filta-Max™/Puri-Max™ system is very efficient at entrapping *Cryptosporidium* oocysts from water samples, and purifying them from the accompanying debris to enable an accurate count to be taken. The system allows simple and efficient recovery of *Cryptosporidium* oocysts from 100-2000 litre volume water samples containing numbers of oocysts ranging from 80-3000. Recoveries averaging 80% were repeatedly observed.

A total of forty QMS™ versus manual comparative counts were performed using various sample slides, ranging from PBS *Cryptosporidium* oocyst spikes to 1000 litre spiked processed water samples. Nine of these of these comparative counts are shown in table 2. A statistical analysis (paired-sample test,[5]) was performed on the data, and showed no statistical difference between manual and QMS counts (p=0.10). QMS™ was then used to analyse five of the slides analysed manually after complete Filta-Max™/Puri-Max™ runs (Table 1), giving comparable counts.

The most difficult samples are those with a high sample volume, which in many cases can generate substantial particulate matter. This, combined with the need to detect low numbers of *Cryptosporidium* oocysts can be particularly challenging.

Table 1 shows data from 20 runs that were performed using high sample volumes (of which 16 were 1000 litres or more), five of which had concentrate turbidities in excess of 1000 nephelometric turbidity units (n.t.u). These five samples had been spiked with low numbers of *Cryptosporidium* oocysts (234 oocysts) and recoveries of 33-44 % were still achieved.

Table 1   *Recovery of low cryptosporidium oocyst spikes from large volume water samples, using the combined Filta-Max™/Puri-Max™/QMS™ System*

Volume filtered (l)	Turbidity (n.t.u.)	Oocyst Spike	Eluate Recovery Manual	QMS™
1000	50	70	76%	
1000	100	80	95%	
1000	240	150	59%	nd*
1000	400	80	45%	nd
1000	400	160	68%	nd
1046	950	234	40%	nd
2000	2070	234	34%	nd
1000	nd	120	48%	nd
200	nd	1500	61%	nd
500	nd	500	78%	nd
1000	nd	3000	99%	nd
500	nd	250	94%	nd
1000	nd	1150	91%	nd
1000	nd	500	94%	nd
1000	nd	120	66%	nd
979	1220	234	37%	37%
1055	860	234	27%	27%
1273	7500	234	44%	38%
1497	5000	234	37%	34%
1324	6250	234	33%	35%

* nd = not determined

Table 2   *A comparison of manual counts with QMS™ counts*

Sample Number	QMS Count	Manual Count
1	56	50
2	79	77
3	35	33
4	85	88
5	29	27
6	27	27
7	29	38
8	24	23
9	79	86

6 DISCUSSION

The Filta-Max™ system has reliably reproduced recoveries of over 80 % from a range of sample types from raw water to finished drinking water[8], with oocyst spikes as low as 80, and has successfully handled sample through-put of 100-1800 litres[6]. Puri-Max™ recoveries have also reliably been in the 80+ % region for finished waters and 60+ % for raw waters[8]. The data suggests that these results are independent of sample type and volume or the level of oocyst challenge used. The QMS™ has shown a performance level comparable or better than manual counting, with no statistical difference between the two. The QMS™ counts, are potentially more accurate than manual because the predefined scan pattern prevents the overlapping or omission of fields, so preventing over/under counting. In addition, reviewing individual features at 100X with a high quality optical/camera configuration reduces misclassification. QMS™ users will experience less fatigue during the scanning procedure leading to greater accuracy. The ease with which features can be retrieved from previously scanned slides is particularly important for external verification of oocyst counts. There is no requirement for data to be recorded by-hand, all data/images are automatically stored and can be printed at the end of the review. QMS™ provides full traceability of sample identity, as the slide label image is automatically stored. There is an option to record on videocassette the entire scanning and review sequence, so the QMS™ user has an indisputable record of the enumeration procedure.

The combined system is inexpensive to purchase, simple to use and provides rapid, reliable recoveries, plus a fully traceable record of the analysis. Sample reference; feature position, operator details; and video traceability are all available if required. This development represents a significant step forward in terms of simplifying sample concentration, extraction, clean-up and analysis to enable the generation of reproducible data for *Cryptosporidium* presence in water.

REFERENCES

1. D. J. Dawson, M. Maddocks, J. Roberts, and J. S. Vidler, *Letters in Applied Microbiology*, 1993, **17**, 276.
2. W. R. MacKenzie, N. J. Hoxie, M. E. Procter, et al., *New England Journal of Medicine*, 1994, **331**, 161.
3. C. E. Musial, M. J. Arrowood, C. R. Stirling, and C. P. Gerba, *Applied and Environmental Microbiology*. 1987, **53**, 687.
4. J. E. Ongerth, and H. H. Stibbs, H.H, *Applied and Environmental Microbiology, 1987*, 1987, **53**, 672.
5. R. E. Parker, *Introductory statistics for biology studies in Biology No. 43.*
6. A. Parton, F. Mendez, and D. P. Sartory, *International Association on Water Quality*, 1997, 185.
7. A. J. Richardson, R. A. Frankenberg, A. C. Buck, *Epidemiology and Infection*, 1991, **107**, 509.
8. D. P. Sartory, D.P., A. Parton, A. C. Parton, J. Roberts, and K. Bergmann, *Letters in Applied Microbiology*, 1998, **27**, 318.
9. K. M. Shepherd, and A. P. Wyn-Jones, *Water Science and Technology*, 1995, **31**, 425.

10. H. V. Smith, L. J. Robertson, and A. T. Campbell, *European Microbiology*, 1993, **2**, 22.
11. G. Vesey, J. S. Slade, M. Byrne, K. Shepherd, and C. R. Fricker, *Journal of Applied Bacteriology*, 1993, **75**, 82.
12. T. N. Whitmore, and E. G. Carrington, *Report FR 0274, Foundation for Water Research, Marlow, UK*, 1992.

THE EXPERIENCE OF THE LEAP PROFICIENCY SCHEME WITH RESPECT TO *CRYPTOSPORIDIUM* TESTING

K. Clive Thompson   Barry May   Diane Corscadden and John Watkins
Alcontrol Laboratories LEAP-CSL Scheme   CREH *Analytical* Limited
Rotherham S60 1BZ   Sand Hutton   Horsforth
                       York YO41 1LZ   Leeds, LS18 4RS

1 INTRODUCTION

Internal and external quality schemes are an important part of a laboratory's analytical programme. In addition they form a basic requirement for accreditation schemes such as the United Kingdom Accreditation Service (UKAS). In the past, trials which have tested the ability of laboratories to recover *Cryptosporidium* and *Giardia* from samples have demonstrated that the claims made do not always match the results obtained.[1] On this particular occasion, samples seeded with *Cryptosporidium* oocysts and *Giardia* cysts as well as false positive samples were sent to 16 laboratories for analysis. Six laboratories were unable to recover the parasites. The recovery rate for the remaining laboratories was between 1.3 – 5.5% for *Cryptosporidium* and 0.8% - 22.3% for *Giardia*. The mean value for *Cryptosporidium* was 2.8% and for *Giardia* was 9.1%. False positive results were also reported.

In 1974, when Yorkshire Water was formed, a monthly internal quality assurance scheme was established ('The AQC Scheme'). The scheme originally provided chemical samples but was widened to provide bacteriological samples in addition. In 1992, this scheme became the Laboratory Environmental Analytical Proficiency (LEAP) Scheme. Since then the scheme has broadened its activities to provide a chemical and microbiological proficiency scheme for the water industry in the United Kingdom and more recently, overseas. The activities of the scheme have been designed to reflect the requirements of regulatory bodies overseeing analysis within the industry. The scheme is now organised by The Central Science Laboratory, an executive agency of the Ministry of Agriculture, Fisheries and Food at York.

2 OPERATION OF THE LEAP SCHEME

2.1 External Quality Assurance

The main purposes of a proficiency scheme may be defined as:-

- The comparison of an individual laboratory's results with an external standard of quality.
- The comparison of results with peer laboratories.

- The comparison of current results with past performance.
- The comparison of new and old methods.
- The assessment of new analysts who have been trained in analytical techniques.
- To provide analysts with confidence in the quality of their data and their analytical technique.

The main objectives of proficiency testing are firstly to provide a regular, objective and independent assessment of the accuracy of a laboratory's results when analysing routine samples. In addition, the information gained should help the laboratory to improve the accuracy of routine analytical data.

2.2 Microbiological Test Samples Offered by the LEAP scheme

The first test samples to be incorporated in an internal quality assurance scheme contained coliforms and *Escherichia coli*. These were introduced routinely in 1987. Enterococci and plate counts were included in 1993, organism biotyping in 1995 and *Clostridium perfringens* in 1997. Parasitology in the form *Cryptosporidium* and *Giardia* were included in 1993. All LEAP bacteriology samples are full volume samples which require no pre-treatment or dilution prior to analysis.

The LEAP Scheme currently offers four types of sample for parasitology. The first of these is a tap water concentrate containing both *Cryptosporidium* and *Giardia*. This is designed to assess the ability of an analyst to correctly prepare slides, fix and stain them and count the bodies. The second sample is a wound polypropylene (Cuno®) filter seeded with *Cryptosporidium* oocysts. The third sample, containing both parasites, is a suspension which is added to 10 litres of tap water. This allows participating laboratories to recover the organisms using their own routine test procedure. The fourth sample is a Genera Filta-Max™ filter spiked with both parasites. Participating laboratories are invited to an annual meeting to discuss the results of the previous year. Suggestions are also invited for improving the scheme and the associated analysis.

The LEAP Scheme has provided six sets of *Cryptosporidium* samples per year since 1993. The results from these exercises have clearly shown a definite improvement over time for this difficult parameter. It is generally found in proficiency schemes that the quality of results for "difficult determinations" improves for a group of laboratories working together. Evidence of this can be seen by comparing the results in table 2 and Table 3.

Oocysts for the Scheme are purchased from Moredun Scientific Limited as live suspensions in phosphate buffered saline (PBS). Suspensions are purchased on the regular basis of four times a year as new batches of oocysts are produced. They are heat inactivated prior to preparation and distribution of the samples (see below). *Giardia* cysts are currently purchased from Waterborne Inc. USA as suspensions in 5% v/v formalin. Suspensions are kept at $2 - 8°C$. Dilutions for the samples are prepared in PBS 5 – 6 days before the samples are shipped. The suspensions are checked to ensure that they have the appropriate levels of parasites in them. The test samples are usually seeded five days before shipment and stored at $2 - 8°C$. Samples are shipped overnight by courier and analysis requested to complete analysis within two weeks. No attempt is made in the data collected to determine whether one or more analysts are involved with the samples. Instructions are given, however, that analysts prepare and count their own slides for the sample provided specifically for counting. Each test suspension is counted by the LEAP Scheme approximately one week after the samples have been shipped. When the Scheme first started, counts were made using 4 x 10µl of each suspension. As numbers of parasites have been reduced, counts are now prepared using either 10 x 20µl or where numbers are very low 10 x 50µl. Recoveries are therefore based on counts of the actual suspensions used to seed all the samples.

The scheme started by adding large numbers (e.g. 5,000-20,000) of parasites to the samples which require recovery. This was designed to give the laboratories every opportunity to recover the parasites and generate confidence in the techniques within the staff. During the last three years levels have reduced substantially to those which might be expected to be found in "real" samples. Every effort is now made to seed the samples at approximately 50-200 parasites per sample.

The parasites used in the scheme are non-viable to ensure safe transportation. The *Cryptosporidium* oocysts are heat-killed at 60°C for five minutes just prior to sample preparation and the *Giardia* cysts are fixed in formalin. Recovery of the parasites does not appear to be affected by rendering them non-viable. In addition, it does not affect staining or identification of internal contents using 4',6-diamidino-2-phenylindole (DAPI).

The scheme has been running *Cryptosporidium* and *Giardia* test parameters for eight years. Early results from the scheme are given in a previous publication[2]. The purpose of this presentation is to provide an update of the lessons learned since then. Participating laboratories have been randomised and the lettering system is not consistent throughout the Tables to prevent identification of laboratories.

2.3 A Chemists View of Detection Limits

The current regulatory method for the detection of *Cryptosporidium* in treated waters requires the analysis of 1,000 litre samples taken over a 24 hour period[3]. Recovery exercises in laboratories use a seed of 100 oocysts. Assuming that a true detection limit of three oocysts in 1,000 litres of water is required, this would correspond to one oocyst based on the minimum regulatory recovery efficiency of 30% for the test procedure. This is equivalent to an approximate oocyst mass in the 1,000 litres of approximately 200 pg (200×10^{-12}g). When this is expressed as a concentration in the original 1,000 litres of water, it is equivalent to 0.2 pg/litre (0.0002 ng/litre). The requirement for a detection limit as small as this would seem daunting. It is worth noting that the most stringent regulatory limit for a chemical parameter in the current water supply regulations is for benzo-(a)-pyrene. This parameter has a maximum permitted concentration of 10ng/litre. In order to monitor this, a detection limit of 1ng/litre is required. This limit is 5,000 times less stringent than that required for regulatory *Cryptosporidium* analysis.

2.4 Suspension for Counting

This part of the scheme was validated at the beginning by preparing replicates of a single suspension and counting this over a period of 27 days. Table 1 shows that although a single suspension was used, the replicates prepared from it give very close mean values over the test period.

Table 1 *Replicates of a Single Suspension Tested over a Period*

Replicate	1	2	3	4	5	6	7	8	9	10	Mean
Day											
0	10	10	11	18	10	7	16	8	12	13	11.5
2	13	9	7	13	13	9	7	15	10	8	10.4
4	13	16	13	9	8	8	13	11	13	12	11.6
7	7	11	9	11	12	10	11	15	12	12	11.0
9	9	13	15	12	12	9	7	11	17	11	11.6
13	15	10	15	14	13	8	15	12	9	14	12.5
17	7	8	15	7	8	10	11	10	12	11	9.9
21	16	10	7	11	8	14	7	12	10	12	10.7
27	10	9	8	19	16	11	14	2	16	17	12.2
Mean	11.1	10.7	11.1	12.7	11.1	9.6	11.2	10.7	12.3	12.2	11.3

The grand mean for this data is 11.3. The data shows that viable oocysts used for counting purposes are stable over the test period when suspended in phosphate buffered saline (PBS) and stored at $2 - 8°C$. The data was subjected to statistical analysis by Dr Hilary Tillett. Her conclusions were that the data demonstrated a successful preparation where the aliquots give counts of oocysts which vary around the mean value with random variation. In addition, there is no excess dispersion over time and between replicates. Statistical analysis of similar data[4,5] has shown that the data is normally distributed. This data might therefore be compared with the data presented in Table 2 which is a suspension distributed to laboratories for counting.

Table 2 *Data Collected from an Early LEAP Counting Exercise*

Laboratory Number	Oocyst Count in 10µl				Mean
A	1	0	0	0	0.25
B1*	18	15	11	14	14.5
B2*	24	30	17	16	21.8
C	5	7	6	5	5.8
D1	11	2	0	1	3.5
D2	4	3	3	8	4.5
E	0	0	0	0	0
F	0	0	0	0	0
G1	14	2	2	5	6.5
G2	8	0	0	2	4.0
H1	24	14	14	11	16.5
H2	37	35	35	19	27.8
H3	14	17	17	13	15.3
J1	17	18	18	13	15.3
J2	14	13	13	14	14.5
K	10	6	6	10	9.0

* Denotes two separate analysts in the same laboratory.

The mean value as determined by LEAP was 18.5 oocysts in 10μl. The data shows that only a small number of laboratories were close to the mean and two analysts failed to observe any oocysts. In comparison with the last exercise in 2000 (Table 3), the ability to count a suspension has improved significantly.

Table 3 *Replicate Counts from a Recent Exercise*

Laboratory Number	Oocyst Count										Vol/ Well	Mean 10μl
A	34	23	15	33	28	26	31	18	20	25	25μl	10.5
B	18	18	27	18	23	20	13	16	26	20	25μl	8.0
C	19	26	16	11	19	24	20	23	24	22	25μl	8.2
D				3	3	7	10				10μl	5.7
E				4	2	3	10				10μl	4.7
F				9	7	4	5				10μl	6.2
G		2	7	20	6	1	7	9	2		10μl	6.7
H				13	10	8	12				10μl	10.7
I	5	10	22	3	14	8	7	6	22	14	10μl	11.1
J	11	7	13	10	17	9	8	21	13	8	10μl	11.7
K				1	2	8	6				10μl	4.2
L				10	6	9	4				10μl	7.2
M	7	9	12	9	9	3	5	8	4	6	20μl	3.6
N	2	11	3	9	4	5	10	8	3	2	20μl	2.8
O				5	6	14	7				10μl	8.0
P				16	11	8	8				10μl	10.7
Q	16	12	24	21	11	12	13	17	15	13	20μl	7.7

The overall sample mean for the whole distribution of samples was 7.7 oocysts in 10μl and the LEAP mean value based on 10 x 20μl replicates was 9.3. Laboratories have developed a significantly improved ability in counting suspensions and this is clearly where the value of such a scheme lies. None of the laboratories failed to detect oocysts. The volumes used for counting vary and the data would suggest that using only four replicates and 10μl volumes tends to give a significantly reduced count when compared to the LEAP Scheme Results. However, laboratories M and N have reduced counts despite using 10 x 20μl replicates.

The early data (Table 2) prompted the LEAP Scheme to send out pre-stained and counted slides. The results from this exercise demonstrated that all the laboratories who took part were able to count oocysts. The count data returned to the Scheme was close to the value of the pre-counted slides for each laboratory. This would suggest that it is the process of staining rather than counting which gave the poor results in Table 2. It would also suggest that where counting is an important element of analysis, asking analysts to perform the whole staining procedure is a much better assessment of ability than counting alone. The Scheme has not yet tried to introduce non-oocyst bodies which mimic *Cryptosporidium*. This will be tried in the near future.

2.5 Wound Polypropylene Filters

The Scheme also provides a seeded wound polypropylene (Cuno®) filter. This type of filter was originally proposed as the test method for water analysis[6].

Table 4 *Oocyst Percentage* Recoveries from Polypropylene Filters in 1995 and 1999*

Exercise	C1 1995	C2 1995	C3 1995	C4 1995	C5 1995	C6 1995
No. of oocysts	27250	7800	3400	24000	10000	1800
Mean (%)	34.2	13.3	21.9	16.7	14.3	18.5
Median (%)	22.4	5.3	5.3	7.5	5.7	8.9
Recovery range (%)	1.5-102	0-63	0.7-96	0.3-50	0-56	0-83

Exercise	C1 1999	C2 1999	C3 1999	C4 1999	C5 1999	C6 1999
No of oocysts	280	165	310	104	480	845
Mean (%)	13.9	38.7	23.8	34.8	21.7	17.8
Median(%)	9.8	30.3	16.1	35.6	21.1	13.0
Recovery range (%)	3.2-36	5.3-97	0-57	0-82	6.2-44.8	4.7-40.2

*Statistical outliers excluded from calculations.

Table 4 summarises the percentage recovery for the Cuno® filter, comparing data from 1995 and 1999. The number of oocysts used to seed the filter has been significantly reduced. It can be seen that there is a wide range of recoveries across laboratories and between exercises. Both the mean and median of the recoveries were low but some improvements can be seen in 1999 as the median values are closer to the mean value, suggesting that a greater proportion of the laboratories are recovering oocysts. There are however some laboratories (even in 2000) who have difficulty in recovering any oocysts from these filters. The principal problem with this type of filter is the large volume of filter washings that require further processing. Large volume centrifugation of the washings will give poor recoveries and this is perhaps the single most important oocyst loss step. Laboratories which have changed from centrifugation to filtration to concentrate oocysts from filter washings have seen much improved recoveries.

The data in Tables 5 and 6 shows that although the seed levels have been reduced significantly between 1994 and 2,000, laboratories who participate in the scheme have improved their recoveries. Over the last three exercises, we have deliberately reduced the seed concentration to below 100, yet even here, there is only one negative recovery. However, the recovery for this type of filter is still very variable. Laboratory D is a good example of this whereas laboratory G would appear to have the most consistent recoveries. As might be expected with this type of exercise, recoveries exceed 100% once the seed level lowers. The data would suggest that if the recovery protocol can be optimised, recoveries with this type of filter should be in excess of 50%.

Table 5 *Recoveries from Polypropylene Filters for 1994*

Laboratory Reference	Percentage Recovery from Polypropylene Filters					
	C1/94	C2/94	C3/94	C4/94	C5/94	C6/94
A	37	14.5	9.3	58	0	61
B	1.8	3.2	9.4	9.0		2.5
C	7	<0.1	0.2	4.3		2.7
D	<0.1	4.6	0.2	<0.1	0	<0.07
E	24	<0.1	0.9	27	1 oocyst	
F	5.4	2.1	4.2	1.9	0	7.3
G	1.1	0.3	2.3	2.8	0	17.9
H	2.2	<0.1	0.7	3.0	0	11.1
Seed Concentration	9,937	4,300	5,400	5,000	*	6,500

* This sample was seeded with a reservoir water diluted 1:5 with distilled water in an attempt to provide a sample containing *Cryptosporidium*-like bodies such as algae.

Table 6 *Recoveries from Polypropylene Filters for 2000*

Laboratory Reference	Percentage Recovery from Polypropylene Filters					
	C1/00	C2/00	C3/00	C4/00	C5/00	C6/00
A	43.7	85.7	60.0	<10	116.3	123.1
B	28.5	71.4	6.0	92.6	34.9	11.8
C			30.0			
D	1.0	57.1	<0.5	83.3		23.1
E	4.9	82.1	6.0	120.4	58.1	41.1
F	9.9	178.6	10.0	55.5		7.7
G		141.4	79.5	44.4	63.9	61.5
H	<0.12	38.7	9.2	37.0	30.2	24.9
Seed Concentration	755	140	250	54	43	65

2.6 Suspensions for Recovery

Suspensions are provided as 1 ml in PBS. This is added to 10 litres of an appropriate final water and the sample processed by laboratories in the way that they would normally analyse a 10 litre 'grab' sample. The processing includes calcium carbonate flocculation[7] or membrane filtration[8], usually using cellulose acetate membranes. For the cleaning step, where used, the majority of laboratories use immunomagnetic separation. A wide range of centrifugation speeds are used to remove oocysts from the filter washings, from 1100g to 2,400g suggesting that different machines have been shown to give different recoveries. During earlier exercises, sucrose flotation would have been used as the cleaning step.

Table 7 compares recoveries of *Cryptosporidium* from the suspension for the years 1995 and 1999. The mean and the median are closer together indicating a more even spread of recoveries for the suspension. The mean recovery generally is more consistent in 1999, although there are still some laboratories who are unable to recover any oocysts from the suspensions.

Table 7 *Oocyst Percentage Recoveries* from Suspension in 1995 and 1999*

Exercise	C1 1995	C2 1995	C3 1995	C4 1995	C5 1995	C6 1995
No of oocysts	400	7800	800	400	5700	400
Mean (%)	57.4	17.1	22.4	29.7	27.5	53.9
Median (%)	57.4	16.3	27.5	1.5	33.1	62.6
Recovery range (%)	19-96	1.9-34	3.1-37	0-88	2.2-47	0-91

Exercise	C1 1999	C2 1999	C3 1999	C4 1999	C5 1999	C6 1999
No of oocysts	120	260	205	200	410	1080
Mean (%)	41.9	46.1	57.5	34.5	53.4	38.6
Median (%)	48.4	35.6	63.2	26.5	46.15	36.1
Recovery range (%)	0-100	0.4-136	1.0-107	6.6-110	2.4-117	11-81.8

*Statistical outliers excluded from calculations

Table 8 *Recoveries of oocysts from suspensions 2000*

Laboratory Reference	Percentage Recovery from Suspensions					
	C1/00	C2/00	C3/00	C4/00	C5/00	C6/00
A	38.7	78.4	28.1	4.7	<35%	20.5
B	40.3	75.3	5.9	44.3	35.7	
C	37.1	156.9	9.4	0.8		11.3
D	11.3	127.0	3.1	21.9	142.8	46.1
E	21.6	49.8	62.5	83.3	46.4	36.4
F	1.6	0.4	54.4	41.2	21.4	12.8
G	22.6	43.1	68.7	9.4		<5.1
H	36.0	50.6	45.9	50.1	21.4	21.0
I	39.3	101.2	114.4	56.6	46.4	30.2
J	11.0	<0.4	49.7	30.0	125.0	
K	8.9	<0.4	29.1	375.0	0	
L	0.5	19.6	100.0	15.6	6.4	<0.5
M		27.4	58.1	9.7	118.9	41.0
N	24.0	9.8	53.7	21.1	71.4	35.9
O	28.2	12.5	15.0	61.6	57.1	19.0
Seed Concentration	620	255	320	640	28	195

The Scheme has recently tried to reduce the seed concentration in the suspension. The recoveries obtained with the suspensions containing fewer parasites also shows a substantial amount of variability (Table 8). Laboratory I has probably the best recovery record over the year whilst laboratory L has significant variation. Although the majority of laboratories are able to achieve good recoveries there is often poor consistency over the six exercises in Table 8.

2.6 Seeded Genera Filters

Seeded Genera Filta-Max™ filters were introduced into the scheme in 1999. The filters are seeded with approximately 100 oocysts. These filters are analysed in accordance with the Standard Operating Protocol (SOP) as part of new Government regulations [3]. A set of recovery data similar to that for the suspensions is presented in Table 9.

Table 9 *Recovery of Oocysts from Filta-Max™ Filters during 2000*

Laboratory Reference	Percentage Recovery from Genera Filta-Max™ Filters					
	C1/00	C2/00	C3/00	C4/00	C5/00	C6/00
A	29.7	12.7	23.6	32.7	0*	25.6
B	57.1	19.0	6.2	43.3		40.7
C	30.8	41.3	44.0	40.4	3*	18.6
D	42.8	49.2	34.5	43.3	1*	36.0
E	19.8	46.0	48.8	42.3	1*	
F	27.5	66.7	37.4	30.8	24*	30.2
G	33.5	65.1	22.5	8.6	27*	18.6
H		41.3				
I	19.8					
J	38.5	80.9	15.3	52.9	25*	
K			5.4	11.5	15*	40.7
L	28.6	66.7	13.8	15.4	0*	
M		76.2	46.5	33.6	6*	34.9
M	30.8		22.9	24.0	0*	47.7
O	47.2	41.3		17.3	3*	30.2
Seed	92	63	275	104	10	86

* The seed was considered to be too low to calculate percentage recoveries. The data is presented as the actual number of oocysts recovered.

The data here also shows some variation but not as much as with the polypropylene filters or the suspensions. Some laboratories would appear to obtain a reasonably good recovery consistency, for example, laboratory D whilst others, for example, J and L are much less consistent. There have been no zero recoveries during this year with the exception of C5/00. Twenty four of the 59 recoveries (40%) exceed 40% recovery.

The C5/00 exercise was an attempt to get the seed value significantly below 100. The actual value was not intended and we were surprised at the number of laboratories reporting oocysts detected.

3 DISCUSSION

In order for a laboratory to gain maximum benefit from participating in an external quality assurance scheme, it needs to take a long-term view of the quality of data produced. It is not reasonable to examine results in isolation and this is particularly so for microbiological determinands which do not always conform to the simple statistical models often seen in the chemical analysis of water. It is not reasonable to examine individual results in isolation. Mistakes can sometimes be made and occasional low recoveries are to be expected on a statistical basis. The longer-term view is that the majority of laboratories taking part in the scheme have, over the years, improved their ability to count and recover oocysts from prepared samples. Suspension counts, whilst still having some variability, have improved considerably. We seldom see a zero count on a suspension although it does occasionally happen. Clearly those laboratories who achieve consistent low counts have a problem with this part of the analysis. They cannot expect that they will have a great deal of success in tackling a sample if their suspension counts are low.

Figure 1 Cryptosporidium *% of Mean Results Positive of Pooled Mean C1 to C6 for 1996 and 1997Using Suspensions for Counting*

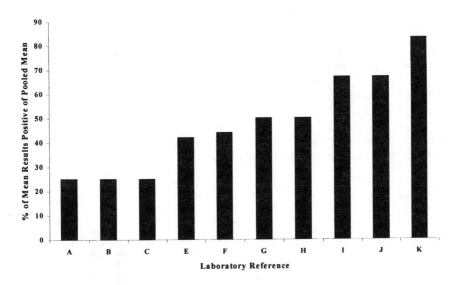

Figure 1 shows the percentage of results from suspensions used only for counting which were positive of the pooled mean over a 2 year period from 1996 to 1997. The mean percentage recovery data for the polypropylene filter for the same period has been calculated. The data is presented in Figure 2.

In general terms, those laboratories which achieve a low percentage of mean results positive of the mean on suspensions provided for counting, also achieve a lower mean percentage on recovery from the seeded filter.

Figure 2 Cryptosporidium *Percentage Recoveries for Polypropylene Filter Exercises C1 to C6 1996 and 1997.*

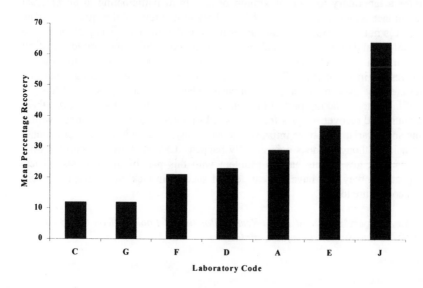

Figures 3 and 4 examine similar data from 1999 and 2000. In these figures there is also a range of percentage mean results positive of the pooled mean and those laboratories which achieve low percentage positive results also have lower mean percentage recoveries when processing suspensions.

Figure 3 Cryptosporidium *% of Mean Results Positive of Pooled Mean C1 to C6 for 1999 and C1 to C4 2000 Using Suspensions for Counting*

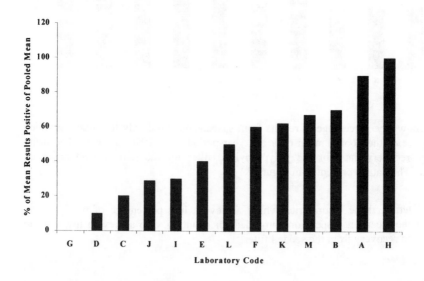

Figure 4 Cryptosporidium *Percentage Recoveries for Suspension Exercises C1 to C6 1999 and C1 to C4 2000.*

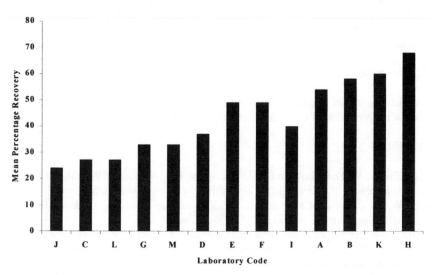

Whilst improving counts on suspensions will not solve all the problems, clearly improving staining and counting will lead to higher recoveries. Laboratories who achieve good results in their counts also achieve good and consistent results in their recoveries. Looking at data over a period of time helps a laboratory to assess and improve its performance. Laboratories such as laboratory H have the best performance in both samples.

Over the years with the seeded samples, the numbers of oocysts have been reduced to make samples more realistic of environmental samples. The early samples used between 5,000 – 20,000 oocysts for each exercise. The numbers are now down to less than 100. Recoveries with the polypropylene filters have improved dramatically as the seed levels have fallen. The technique is still variable, possibly because of the difficulty of washing oocysts out of the filters and perhaps the use of large volume centrifugation. Early experiments suggested that the polypropylene filters can permit up to 30% of oocysts to pass through and therefore dosing water samples and pumping these through the filters was not an alternative that could be used.

Recoveries from suspensions have improved also as the numbers have been reduced. There is however considerable variation here also. Recoveries from the Filta-Max™ filters are less variable and in general recoveries are better than those observed with the polypropylene filters and the suspensions.

The Scheme has reduced the seed levels to the point where counting each seed suspension requires a substantial number of replicates in order to obtain a reasonably accurate estimate of seed levels. To date few samples have been supplied with no oocysts in them. It could be argued that asking analysts to spend time on samples which contain nothing is non-productive. For the future, the scheme will look firstly to the introduction of relevant particulate material into samples, small amounts at first but then increasing these amounts to generate significant pellet volumes. In addition, we are hoping to introduce bodies which resemble oocysts as false positive samples e.g. certain algal cells. It will be interesting to observe how participants respond.

The scheme is intended to help laboratories and analysts to have confidence in the isolation and counting of *Cryptosporidium* oocysts and *Giardia* cysts. Making that scheme ever more challenging is an objective for 2001. We believe that by introducing these changes on a gradual basis, it gives analysts the opportunity to learn more about the techniques that they are using and to gain confidence in their own ability and judgement.

4   ACKNOWLEDGEMENTS

The Scheme would like to thank all the participants for their support over the years and for the provision of the data which makes a comparison of this type possible.

5   REFERENCES

1. J. L. Clancy, W. D. Golnitz and Z. Tabib, *J. Am. Wat. Wks. Ass.* 1994, **86,** 89.
2. K. C. Thompson, B, May and D. Corscadden, in *Protozoan Parasites and Water*, eds. W. B. Betts, D. Casemore, C. Fricker, H. Smith and J. Watkins, Royal Society of Chemistry, Cambridge, 1995, Part 4, p.176.
3. Drinking Water Inspectorate, *Standard Operating Protocol for the Monitoring of* Cryptosporidium *Oocysts in Treated Water Supplies*, Part 2 – Laboratory and Analytical Procedures, November 1999.
4. Drinking Water Inspectorate, *Continuous Monitoring for* Cryptosporidium *in Drinking Water Supplies*, 1999.
5. C. A. Francis, D. Corscadden, J. Watkins and M. Wyer, in Cryptosporidium, *The Analytical Challenge.* This Publication.
6. Anon, 1989, *Detection of* Cryptosporidium *Oocysts and* Giardia *Cysts in Water and Related Materials.* HMSO, London.
7. G. Vesey, J. S. Slade, M. Byrne, K. Shepherd and C. R. Fricker, *J. Appl. Bacteriol.*,1993, **75,** 86.
8. D. J. Dawson, M. Maddocks, J. Roberts and J. S. Vidler, *Lett. Appl. Microbiol.*, 1993, **17,** 276.

AN EVALUATION OF THE CURRENT METHODS FOR THE DETECTION AND
ENUMERATION OF *CRYPTOSPORIDIUM* IN WATER

Carol Francis, Diane Corscadden, John Watkins and Mark Wyer

CREH *Analytical* Limited
Hoyland House
50, Back Lane
Horsforth, Leeds LS18 4RS

CREH
University of Wales
Lampeter
Ceredigion SA48 7ED

1   INTRODUCTION

The analysis of water samples for *Cryptosporidium* and *Giardia* has been described as
insensitive, time consuming, labour intensive and expensive. Added to this is the fact
that when oocysts are detected there is no reliable way to determine whether they are
viable, and more importantly, infective. Unfortunately there are no reliable surrogate
indicators for the presence of *Cryptosporidium* or *Giardia* in water samples[1] and
therefore the definitive way of establishing presence or absence is by analysis of water
samples for the parasites. Risk assessment of catchments and maintenance of good
water treatment practices will help to minimise the presence of these parasites in
drinking water.

Much has been done since the publication of the first UK method for the analysis
of water samples[2] to improve the sensitivity of the methods of analysis. These
advances have been recognised in a second UK method book[3]. The original wound
polypropylene filter has been shown to permit up to 30% of oocysts used in seed waters
to pass through (Watkins, unpublished data)[4]. The introduction of flow cytometry[5,6]
improved the speed at which stained samples could be analysed and improved detection
in samples containing considerable amounts of particulate material. It also enabled the
use of small volume 'grab' samples which could be processed by calcium carbonate
flocculation[7] or by membrane filtration[8,9]. Samples were originally cleaned by flotation
on a sucrose solution of specific gravity 1.18[2]. The introduction of immunomagnetic
separation (IMS) for cleaning samples provided an ideal alternative to sucrose flotation
both in terms of recovery and removal of unwanted particulate material.

In 1999, the Government introduced new Regulations for the monitoring of
Cryptosporidium in treated waters. The standard operating protocol (SOP)[10] proposed
in the Regulations has been evaluated by the Public Health Laboratory Service[11] and
found to give recovery rates between 23 to 42% based on a seed of 100 oocysts. The
method involves sampling using the Genera Filta-Max[TM] filter, elution, concentration
of deposits by filtration and centrifugation, cleaning by immunomagnetic separation

(IMS), staining and detection by microscopy. Each of these processes has been examined in some detail by a number of authors[12,13] and been found to give satisfactory results. Methods for the detection of oocysts in water can be broken down into analytical components. Each of these will have inherent losses and each can be improved in terms of recovery and a reduction in time spent doing the analysis. For example, the practices used in the preparation and staining of slides have been examined in some detail[14,15] and where losses can occur and how these losses can be minimised has been demonstrated.

The original purpose of the presentation was to describe the evaluation of new methods or parts of methods which could be used for the regulatory analysis of *Cryptosporidium* in water. At the time little evaluation work had been done and therefore there was little data that could be presented. The presentation therefore examined some of the parts of the analytical procedure to understand where losses could occur and where improvements might be made. The data presented at the conference has been summarised here and more recent data added.

2   REFERENCE OOCYSTS

The reference isolate of oocysts used in the UK for the validation of recovery methods is the Moredun isolate. This isolate, originally from deer, is propagated by infecting 3 day old lambs by Moredun Scientific Limited, Edinburgh. Batches are produced at regular 3 month intervals. Faecal material is collected and the oocysts removed as soon as practicable after shedding. Cleaned oocysts are supplied as a viable suspension in phosphate buffered saline. The suspension is supplied with a certificate which gives details of the batch production and viability. The oocyst suspension is also checked for its ability to stain with monoclonal antibody and with 4',6-diamidino-2-phenylindole (DAPI) and is checked for suitability in its recovery from seeded water samples before it is released for use by the water industry. The viability of the oocysts is usually 95 – 99% when released, falling to approximately 80% after 3 months. The oocysts keep well in phosphate buffered saline and good recoveries can be achieved with suspensions which are up to 3 years old. The recovery of new batches of oocysts from seeded Genera Filta-Max[TM] filters tested before release is between 70 – 80%.

3   COUNTING OOCYSTS

An integral part of the analysis of water samples is the quality control programme which is used to validate the recovery of oocysts and the suitability of materials. This will rely upon the seeding of water samples with a known number of oocysts and attempting to recover as many as possible. The first question in an analytical technique must therefore be how competent is an analyst in preparing suspensions and counting them for use in recovery trials. In practical terms, the concentration of oocysts in a suspension does not need to be accurate. Between 700 and 1,300 per ml is an adequate range from which recoveries can be performed. The more important question is how accurately can the figure be determined. Table 1 provides a range of replicate counts of an oocyst suspension performed on different dates. Each 10 replicates of 20µl has been prepared and stained in an identical manner.

Table 1 *Replicate Counts of a Single Suspension Using 10 x 20µl*

Date	20µl Aliquots										Mean	S.D.
03.08.99	25	21	17	15	27	26	20	24	24	24	23.30	3.954
05.08.99	27	28	23	25	33	29	31	25	23	34	27.80	3.938
16.08.99	21	25	28	28	24	30	20	26	26	24	25.20	3.120
24.08.99	25	26	31	24	28	28	18	34	16	37	26.70	6.516

Analysis of the data for all the results and for the replicate counts on a daily basis show that the data is normally distributed. Although individual counts have quite a wide variation, the means are close and the standard deviations, with the exception of the series on 24.08.00 are very close. In addition, individual 95% confidence intervals for the means based on a pooled standard deviation for the replicates shows that they are all derived from a single population. The 95% confidence interval for the mean value of each series based on the overall pooled standard deviation is ± 3 oocysts of the mean. By performing a sufficient number of replicate counts of a test suspension for seeding, there can be reasonable confidence that a 100µl seed volume will be within ± no more than 20 oocysts of the mean value.

The process was repeated for 100µl aliquots and the results are given in table 2.

Table 2 *Replicate counts of a single suspension using 10 x 100µl*

Date	100µl Aliquots										Mean	S.D.
15.09.99	80	103	90	119	107	95	90	104	90	89	96.70	11.41
16.09.99	96	103	91	94	79	96	103	108	106	97	97.30	8.64
16.09.99	92	111	93	101	113	99	125	119	104	91	104.80	11.82
17.09.99	93	109	102	88	93	90	90	108	90	105	96.80	8.26

In this second series of counts the means and standard deviations are close, the data is also normally distributed and in addition, individual 95% confidence intervals for the means based on a pooled standard deviation for the replicates also shows that they are all derived from a single population. The 95% confidence interval for the mean values is ± 7 oocysts in each series of counts. The data would suggest that providing that a sufficient number of replicates are performed to establish a count, the mean will give a reasonable estimate of the seed for recovery purposes, assuming that the standard deviation is no more than 20% of the mean value. The mean will be more accurate if larger volumes of suspension are used for the count. In practical terms in our laboratory this method of counting is satisfactory for the purpose of routine recovery exercises and for the comparison of new methods against existing methods. Other studies[16] have calculated coefficients of variance for their seeds and derived values of

16.1% for *Giardia* and 15.5% for *Cryptosporidium*. Similar comparative counts have been done[17]. For low seed counts, flow cytometry and Coulter counting were found to have the lowest variability. Standard counting was found to be less accurate and a haemocytometer could not be used to determine low seed values and was only of use when suspension values reached in excess of 6 x 10^4 oocysts/ml.

Suspensions prepared in phosphate buffered saline appear to be stable both in counting and in their recovery from seeded filters over a four week period.

4   RECOVERY OF OOCYSTS FROM WATER SAMPLES

4.1 Concentration

A variety of different methods have been published for the recovery of oocysts from water. These include wound polypropylene filters, flat-bed membrane filters, calcium carbonate flocculation and the Genera Filta-Max™ filters. Wound polypropylene filters will allow oocysts to pass through. Up to 5 – 30% may pass through the filter and 2 – 30% of oocysts seeded onto filters may not be eluted off again[4]. Cellulose acetate membrane filters have been found useful for the recovery of oocysts from small volumes of water (10 – 100 litres)[9]. Recovery from seeded tap water has been found to give a mean recovery of 25.5%. Similar values have been obtained elsewhere[18]. They found recovery from tap water to be between 21.8 – 39.7% using a variety of different membranes including Versapor, cellulose nitrate, cellulose acetate and polycarbonate based membranes. Calcium carbonate flocculation is a simple method to use and has been shown to give good recoveries (over 68% in both seeded tap and river water[7] and over 70% for both seeded tap water and river water[18]). Recovery data produced by the water industry in the UK has also been published[3.] Membrane filtration was examined as a standard method for recovering oocysts from water and the lessons learned used to improve the recovery of oocysts from wound polypropylene filters (see below).

4.2 Centrifugation

Centrifugation is an important step in the concentration of oocysts from water samples. The original method used by the water industry had small and large volume centrifugation in it[2]. Material concentrated from water could be centrifuged as many as 6 times during the recovery process. The recommended speed to be used was 1,500g for 10 minutes. Examination of centrifugation showed that above 500g the numbers of oocysts recovered reduced significantly[4]. At 12,000g only 40% of seeded oocysts were recovered. A similar study[18] found that at 1,500g, only 60% of seeded oocysts were recovered. Recovery improved if the centrifugation speed was increased to 5,000g. In a separate study,[17] the use of higher centrifugation speeds did not lead to diminished recovery of *Cryptosporidium* or *Giardia*. In addition, they found that using centrifuges with the brake off gave only minor improvements in recovery which they considered should be weighed against the extra time allowed for the rotor to stop.

We examined centrifugation in 3 different machines at differing speeds and using different volumes. Oocysts were suspended in 10 litres of tap water and concentrated from this volume using filtration through 142 mm, 1.2μm cellulose acetate membranes (Sartorius). The membranes were eluted in 2 x 25 ml of 0.1% tween 80 in deionised water and the 50 ml eluate centrifuged and counted. Each percentage figure is the mean of 5 replicates. The results are given in table 3. Where large volume centrifugation was used the final volume was made up to 500ml.

Table 3 *Recoveries Obtained Using Different Centrifuges and Centrifuge Volumes*

Centrifuge	Speed (g)	Time (Minutes)	Volume (ml)	Rotor Type	Recovery (%) *C. parvum*	*Giardia*
Machine 1	1,500	10	50	Swing out	115	89
Machine 2	1,500	10	50	Swing out	47	89
Machine 2	2,000	10	50	Swing out	80	84
Machine 2	2,500	10	50	Swing out	87	85
Machine 3	1,500	10	500	Fixed	24	53

Small volume centrifugation in machines 1 and 2 did not affect the recovery of *Giardia*. This is understandable in view of the fact that the parasite is much larger than *Cryptosporidium*. The recovery of *Cryptosporidium* is however dramatically affected. Machine 1 gave excellent recoveries whereas machine 2 gave only 50% recovery using the same time and g force. It was not until the machine speed was increased to 2,500g that suitable recoveries were obtained. The standard method contains at least 5 centrifugation steps. The best recovery one could hope for in machine 2 is 3% unless the speed is increased to 2,500g or the period of centrifugation is increased. Increasing centrifugation speed does improve recovery without apparent losses. However, it would appear that with machine 1, a speed of 1,500g for 10 minutes is more than adequate. Large volume centrifugation gives poor recoveries. The wound polypropylene filter produces volumes of eluate of 2 – 4 litres. This will result in at least 2 large volume centrifugation steps and the best recovery that can be achieved here is 12%. Modifying the removal of oocysts from large volumes of eluate by changing to membrane filtration using cellulose acetate membranes instead of centrifugation will improve the recovery from less than 10% to better than 50% (Watkins – data not shown).

4.3   Cleaning

Filtration and flocculation procedures may recover a substantial amount of particulate material from water in addition to oocysts. The standard method for removing unwanted particulate material has been to use sucrose flotation. At best this gives only 50% recovery in clean samples and 25% or less recovery in turbid samples[3]. Potassium citrate has been used as a replacement for sucrose and Percoll-sucrose has also been evaluated[3,9]. The introduction of immunomagnetic separation (IMS) for the removal of particulate material has greatly improved the recovery of oocysts from water samples. In clean water samples 95% of oocysts can be recovered[13]. The technique has been shown to achieve better than 90% recovery from clean waters and 60% recovery from turbid waters even when oocysts have been 'aged'.[14]

4.4 Staining

Detection of *Cryptosporidium* relies on staining with monoclonal antibody. Earlier methods of staining with a modified Ziehl Neelsen method were found to be too non-specific for environmental samples. The current method of staining is to use a monoclonal antibody conjugated with fluorescein isothiocyanate. The staining reaction can vary, depending upon the condition of the oocysts, the manner in which they have been stored, exposure to chemicals, for example, chlorine or formalin, the time and temperature at which staining takes place and whether the staining is done on a slide or in suspension. The manner in which the slides are handled will also be critical [15,16]. Staining in suspension has been shown to give higher counts than staining on slides[19]. Fixing with 10% formalin can dramatically reduce the stain intensity of oocysts[20]. Stain intensity also decreases with the age of the oocyst and with exposure to chlorine[22,23]. The type of slides, fixative, drying temperature, staining time and washing procedure can reduce oocyst stain intensity and count. A reduction of up to 70% in counts has been observed during the staining process[16].

4.5 Microscopy

The degree of fluorescence observed under a microscope will depend on a variety of factors. These include the running time of the bulb, power of the bulb, correct alignment and focusing of the bulb, cleanliness of objectives and the ability of the microscopists to count oocysts. Table 4 shows the ability of a number of laboratories to count a suspension prepared for the LEAP Scheme (see Thompson *et. al.,* this volume). In addition, the presence of significant amounts of particulate material can influence recoveries. It has been demonstrated that where the turbidity of a suspension is greater that 150 NTU, the stained oocyst count on slides can be reduced by as much as 95%[23]. Cleaning processes must therefore remove as much unwanted particulate material as possible.

Not only do different analysts in the same laboratory get different results but some laboratories get poor results whoever analyses the sample.

Previous work has shown that the distribution of a single suspension into a number of sub-aliquots does not affect the count obtained over a period of 27 days [21].

5   THE EFFECT OF THE ABOVE OBSERVATIONS ON THE CURRENT STANDARD OPERATING PROTOCOL FOR THE RECOVERY OF OOCYSTS FROM WATER SAMPLES

The current Standard Operating Protocol (SOP)[9] for the recovery of *Cryptosporidium* oocysts from drinking water uses the Genera Filta-Max[TM] for the recovery of oocysts from large volumes of water, IMS for cleaning recovered material and staining on slides and direct microscopy for identification and enumeration. Providing that laboratories can count suspensions correctly, have a centrifuge that gives good recoveries (80 – 100%) and follow the protocol as accurately as possible, it should be possible to achieve reasonable recoveries from seeded 10 litre water samples using seeds of between 80 – 120 oocysts. These reasonable recoveries could be defined as being better than 50% on a regular basis.

Table 4 *Counts Obtained by a Number of Laboratories Using 4 x 10µl Replicates of a Single Suspension*

Laboratory Number	Oocyst Count in 10µl				Mean
1	1	0	0	0	0.25
2A	18	15	11	14	14.5
2B	24	30	17	16	21.8
3	5	7	6	5	5.8
4A	11	2	0	1	3.5
4B	4	3	3	8	4.5
5A	0	0	0	0	0
5B	0	0	0	0	0
6A	14	2	2	5	6.5
6B	8	0	0	2	4.0
7A	24	14	14	11	16.5
7B	37	35	35	19	27.8
7C	14	17	17	13	15.3
8A	17	18	18	13	15.3
8B	14	13	13	14	14.5
9A	10	6	6	10	9.0

Figure 1 gives the recoveries obtained in our laboratories over a 12 month period using the above protocol.

Figure 1 *Recoveries of Cryptosporidium and Giardia obtained from seeded tap water samples over a 12 month period*

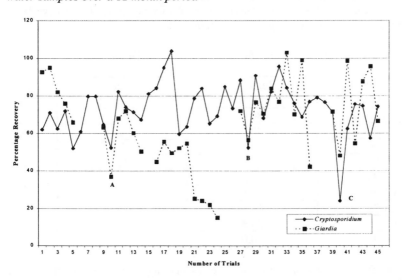

The recoveries show that between 60 – 80% recovery should be achievable on a regular basis. The oocyst source used for this data is the Moredun Scientific isolate. The original source for *Giardia* was a faecal sample obtained from the Liverpool School of Tropical Medicine which was cleaned by ether flotation.

Recovery of *Cryptosporidium* is reasonably consistent at between 60 – 80%. At 3 points labelled A, B and C the recovery of both parasites is reduced. This would suggest an error in executing the test procedure. At point A, an excess of dissociated oocysts was removed from the eppendorf tube. Two slides were required for an approximate volume of 80µl. This would suggest that a small amount of buffer remained in the tube after concentrating the beads. The volume of hydrochloric acid added would not be sufficient to reduce the pH to a level to achieve complete dissociation. At point C the membrane from the Genera wash tube was eluted into 2 centrifuge tubes. The likelihood here is that only 1 tube was processed after centrifugation. The explanation for B is unknown but is almost certain to be an analytical error.

Over a period of time, the *Giardia* recovery deteriorated. This suspension was replaced by a suspension obtained from Waterborne Inc. USA which has performed satisfactorily. During the period of poor *Giardia* recoveries, the recoveries for *Cryptosporidium* were not affected suggesting that the suspension was the cause of the problem rather than the test procedure. The replacement suspension improved recoveries. Suspensions for this figure were prepared on a monthly basis in phosphate buffered saline and stored at 2 – 8°C. These suspensions appear to be perfectly stable. At least 5 different batches of oocysts have been used for recoveries over the year and the method of production does not appear to affect recoveries. The method of counting the suspensions is by using 10 x 20µl aliquots counted on the day that the recovery is performed. This method of enumeration appears to be perfectly satisfactory to achieve a good recovery pattern.

5.1   Age of oocyst suspensions

The process of regular testing of a laboratories ability to recover oocysts inevitably leads to the accumulation of oocyst suspensions of varying ages. These have been tested through the SOP and the results are given in table 5.

Table 5 *Recoveries Obtained Using the SOP on Oocysts of Varying Ages*

Oocyst Batch (Moredun)	Recovery (%)	Age (Months)
C3.00	80.0	1
C2.00	80.3	4
C1.00	92.5	8
C3.99	64.2	12
C2.99	40.2	15
C4.98	55.5	18
C1.98	63.6	30
C3.97	50.3	36
C2.97	60.0	39

The data set shows clearly that although oocysts may age and lose viability over a period of time, the SOP can recover oocysts with better than 50% efficiency even when they are over 3 years old although the best recoveries are achieved with oocysts which are no more than 12 months old.

6 DISCUSSION

The analysis of water and related materials for *Cryptosporidium* is still a time consuming and expensive technique. The sensitivity has improved dramatically over the past 10 years as analysts have improved various stages in the methodology and their own experience. Some things, however may still be taken for granted. The ability to prepare, stain and count suspensions may be poor. Centrifugation speeds and times may not be accurate. Analysis of samples containing large amounts of particulate material may interfere with IMS or mask detection during microscopy. Meticulous attention to detail will help in the reliable recovery of oocysts from seeded samples. Proper calibration and assessment of equipment used in detection will also help. Good training is essential.

There are a number of external quality assurance schemes available. These provide laboratories with samples which may or may not contain oocysts, which may contain interfering material and may contain objects which resemble oocysts. In one such exercise in America[24], of 16 laboratories provided with samples seeded at 800 oocysts, 6 were unable to recover the parasites. The recovery rate for the remaining laboratories was between 1.3 – 5.5%. False positive results were also reported. In addition, laboratories often find that although internally seeded samples give good recoveries, participation in an external quality assurance scheme gives much poorer results. The reasons for this are unclear and much more work needs to be done to ensure that both internal quality assurance and external schemes provide laboratories with material which will give them reliable results. In addition, such schemes should provide laboratories with the confidence to undertake analysis and make decisions about drinking water quality which water utilities can use with confidence.

References

1. E. C. Nieminski, W. D. Bellamy and L. R. Moss, *J. Am. Wat. Wks. Ass.*, 2000, **92**, 87.
2. Anon, 1999, *Detection of* Cryptosporidium *Oocysts and* Giardia *Cysts in Water and Related Materials.* HMSO, London.
3. Anon, 1999, *Isolation and Identification of* Cryptosporidium *Oocysts and* Giardia *Cysts in Water.* Environment Agency.
4. G. Vesey and J. Slade, *Wat. Sci. Tech.*, 1991, **24**, 165.
5. G. Vesey, J. S. Slade, M. Byrne, K Shepherd and C. R. Fricker, in *New Techniques in Food and Beverage Microbiology*, eds. R. G. Croll, A, Gilmour and M. Sussman, Society for Applied Microbiology, Technical Series 31, Blackwell Scientific Publications, 1993, Chapter 8, p. 101.
6. J. Watkins, P. Kemp and K. Shepherd, in *Protozoan Parasites and Water*, eds. W. B. Betts, D. Casemore, C. Fricker, H. Smith and J. Watkins, Royal Society of Chemistry, Cambridge, 199, Part 4, p. 115.
7. G. Vesey, J. S. Slade, M. Byrne, K. Shepherd and C. R. Fricker, *J. Appl. Bacteriol.*, 1993, **75**, 86.
8. J. E. Ongerth and H. H. Stibbs, *Appl. Environ, Microbiol*, **53**, 672.
9. D. J. Dawson, M. Maddocks, J. Roberts and J. S. Vidler, *Lett. Appl. Microbiol.*, 1993, **17**, 276.
10. Drinking Water Inspectorate, *Standard Operating Protocol for the Monitoring of* Cryptosporidium *Oocysts in Treated Water Supplies*, Part 2 – Laboratory and Analytical Procedures, November 1999.

11. Drinking Water Inspectorate, *Continuous Monitoring for* Cryptosporidium *in Drinking Water Supplies*, 1999.
12. D. P. Sartory, A. Parton, A. C. Parton, J. Roberts and K. Bergmann, *Lett. Appl. Microbiol.*, 1998, **27**, 318.
13. A. T. Campbell and H. V. Smith, *Wat. Sci. Tech.*, 1997, **35**, 397.
14. A. T. Campbell, B. Gron and S. E. Johnson, in *1997International Symposium on Waterborne* Cryptosporidium *Proceedings,* eds. C. R. Fricker, J. L. Clancy and P. A. Rochelle, 1997, Newport Beach, California, p. 153.
15. H. V. Smith and C. R. Fricker, in *2nd U. K. Symposium on Health-Related Water Microbiology,* eds. R. Morris and A. Gammie, 1997, p. 170.
16. A. P. Walker, see appropriate chapter in this publication.
17. M. W. LeChevallier, W. D. Norton, J. E. Siegel and M. Abbaszadegan, *Appl. Environ. Micro*, 1995, **61**, 690.
18. F. W. Schaefer III, J. B. Bennett, R. E. Stetler and H. D. A. Lindquist, in *Towards a Standardised Experimental Design for Viability and Inactivation Studies*, Drinking Water Inspectorate, 1999, p.7.
19. K. Shepherd and P. Wyn-Jones, *Appl. Environ. Microbiol.*, 1996, **62**, 1317.
20. R. Hoffman, C. Chauret, J. Standridge and L. Peterson, *J. Am. Wat. Wks. Assoc.*, 1999, **91**, 69.
21. K. C. Thompson, B, May and D. Corscadden, in *Protozoan Parasites and Water*, eds. W. B. Betts, D. Casemore, C. Fricker, H. Smith and J. Watkins, Royal Society of Chemistry, Cambridge, 1995, Part 4, p.176.
22. A. G. Moore, G. Vesey, A. Champion, P. Scandizzo, D. Deere, D. Veal and K. L. Williams, *Int. J. Parasitol.*, 1998, 1205.
23. J. B. Rose, L. K. Landers, K. R. Riley and C. P. Gerba, *Appl. Envirol. Micro.*, 1989, **55**, 3189.
24. J. L. Clancy, W. D. Golnitz and Z. Tabib, *J. Am. Wat. Wks. Ass.* 1994, **86,** 89.

AUTOMATED DETECTION AND VIABILITY ASSESSMENT

D.A. Veal, M.R. Dorsch, and B.C. Ferrari

Department of Biological Sciences,
Macquarie University,
Sydney, NSW 2109, Australia

1 INTRODUCTION

Methodologies for the routine detection of *Cryptosporidium* are in their infancy and can still be best described as modified research protocols. Whilst incremental improvements have and are being made in all steps of the analytical chain (see (1) (2), the high costs of analysis, slow turn around times, variability in recovery rates and low laboratory capacity attest to the cumbersome nature of routine methods currently available. The limitations of these methods become particularly apparent in crises such as that which occurred in Sydney in 1998, when the numbers of samples that needed to be analysed increased by orders of magnitude, stretching the availability of infrastructure and suitably qualified personnel (3). Further, questions concerning the health implications of positive findings have been raised that relate to the species, viability, infectivity of the oocysts and even if oocysts are actually present (4). Currently, no methods suitable for routine detection are available that address these questions. Robust methods for the analysis of *Cryptosporidium* in water are being demanded that minimise the need for subjective and highly skilled microscopy, that provide an indication of the potential health risks, and that in the event of a crisis are capable of analysing large numbers of samples reliably with existing staff and infrastructure. Such detection methodologies are likely to require an increasing level of automation necessitating improvements in both biological reagents and instrumentation.

2 SELECTION OF ANTIBODIES

All the routinely used methods for the detection of *Cryptosporidium* in water rely upon antibody-based techniques. Antibodies are the basis for the selective separation of oocysts and from environmental debris using techniques such as flow cytometry (FCM) and immunomagnetic separation (IMS). Fluorescently labelled antibodies are used in immunofluorescence assays (IFA) to tag oocysts for subsequent identification and enumeration using epifluorescence microscopy.

Despite their pivotal importance in *Cryptosporidium* testing it is perhaps surprising how rarely laboratories consider the influence of antibody quality on the reliability of detection. One study (5) found considerable variation between commercially available antibodies in terms of cross-reactivity and affinity. Of particular concern was the finding of significant differences between batches from some manufacturers. Such differences can result from variations in antibody concentration, purity and from the number of fluorochromes attached to each antibody (F/P ratio).

For environmental analysis non-specific binding of antibodies appears to be one of the critical factors limiting the sensitivity of detection methods. We have developed a flow cytometric assay for determining relative non-specific binding (6) and used it to compare various antibodies, antibody formulations and different water types (Figure 1; (7)). The assay is based on the increase in the numbers of non-target particles detected after mAb staining. The lower the increase in the number of particles the lower the non-specific binding (NSB) ratio and the more suitable the antibody is for analysing that particular sample matrix.

All currently commercially available *Cryptosporidium* mAbs used for detection
- bind to surface epitopes on the oocyst wall.
- are not specific to *Cryptosporidium parvum* oocysts.
- recognise the same surface epitope.
- are either of an IgG_3 or IgM subclass.

Experience gained from clinical diagnostics suggests that antibodies of the IgM or IgG_3 subclass show higher non-specific binding properties resulting in high background fluorescence compared with antibodies of the IgG_1 subclass. Further, IgM and IgG_3 antibodies are more difficult to conjugate to fluorochrome molecules and have lower affinities for their target epitopes than IgG_1 antibodies.

An aim of our research was to develop IgG_1 antibodies against the oocyst wall. However, using whole oocysts we have found it difficult to induce the strong secondary IgG immune response in mice required to produce IgG_1 mAbs. By contrast mice injected with a soluble oocyst extracts showed a strong IgG_1 response (8). The IgG_1 antibodies that have been generated from these mice
- Show reduced binding to debris particles such as algae and minerals in environmental samples, therefore producing low background fluorescence.
- Give bright fluorescence signals when used in IFA.
- Can be easily conjugated to a number of fluorochrome molecules without increases in non-specific antibody binding.
- Significantly increase the sensitivity and recovery of the entire detection system using either IMS or flow cytometry.

Antibodies, no matter how specific, still show some level of non-specificity, either because of non-specific binding or because the epitopes that the antibodies recognise are shared with non-target organisms. In water samples there have been a number of reports showing that *Cryptosporidium* specific antibodies will bind to algae and other protozoa. No currently available antibodies claim specific detection of *C. parvum*, and none provide information on the viability of *Cryptosporidium* oocysts.

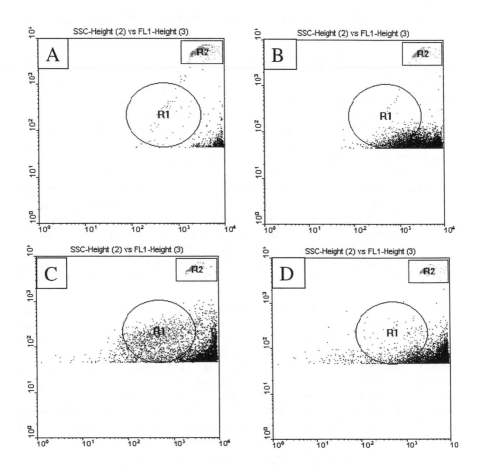

ANTIBODY	NON SPECIFIC BINDING RATIO WATER TYPE	
	Raw	Backwash
Cry 104 (IgG$_1$)	A 0.023± 0.001	B 0.277±0.031
Immucell (IgM)	C 3.294±0.327	D 0.869±0.154

Figure 1 Evaluation of the level of non-specific binding by FITC-labelled Cry 104 and Immucell in raw and backwash water using the method described by (6). The elliptical region (R1) is centred on the position of the main population of pure IFA stained oocysts and a rectangular region is defined around a standard bead population (data not shown). Raw (A & C) and Backwash (B & D) concentrated water samples containing no oocysts stained with Cry 104 (A & B) or Immucell (C & D). The particles within the oocyst region (R1) contain both autofluorescent and non-specifically bound particles. The number of particles within R1 is divided by the number of beads analysed (R2) to produce the non-specific binding ratio (7). Note the higher NSB found with backwash water relative to raw water and higher NSB using Immucell as opposed to Cry 104

3 FLUORESCENT *IN SITU* HYBRIDISATION (FISH) FOR *CRYPTOSPORIDIUM PARVUM*

Fluorescent *in situ* hybridisation (FISH) is a technique that employs fluorescently labelled oligonucleotide probes targeting the ribosomal RNA (rRNA). Portions of rRNA sequences have evolved at different rates and those that can be targeted for FISH labelling range from highly conserved (throughout a phylogenetic domain) to highly variable (up to species or even sub-species specific). Information on these sequences combined with rRNA structural studies give sufficient information for the effective design of selective oligonucleotide probes at various phylogenetic levels. In the case of *Cryptosporidium* we have produced probes that operate at the level of the genus *Cryptosporidium* (probe CRYGEN) or that specifically target *C. parvum* (CRY 1 and CRY 2) (9; Dorsch unpublished). Ribosomal RNA molecules are naturally amplified in viable cells. In the case of *Cryptosporidium* oocysts there are approximately 3.5×10^5 18S rRNA molecules per oocyst (10).The targeted 18S rRNA molecules will bind a high number of probes resulting in a natural amplification of signal and concomitant increase in sensitivity. Ribosomal RNA is a more useful target than mRNA or tRNA because rRNA is more stable, and contains more information, respectively. An additional advantage is that part of the rRNA is also present as single stranded RNA, negating the requirement for a denaturation step.

The rRNA binding specificity of CRY 1 and CRY 2 has been confirmed using both antisense oligonucleotide probes and RNase treatments. The Cry1 probe has been found to hybridise with all samples of *C. parvum* oocysts tested that were capable of excystation at the time of fixation. This included six different batches of oocysts from Moredun Animal Health Ltd (Midlothian, Scotland), 14 samples of *C. parvum* isolates from humans, two bovine isolates from the UK and two bovine isolates from Sydney. The Cry 1 probe did not hybridise to rRNA from two batches of *C. muris* oocysts or a batch of *C. baileyi* oocysts. All batches tested showed a hybridisation signal with a universal eukaryotic FISH probe (Euk), thereby demonstrating that they were viable and that the fixation, permeabilisation and hybridisation conditions were suitable.

Viable cells must contain rRNA. Thus, FISH has the potential to be used as a surrogate of viability. Hybridisation to aged *C. parvum* oocysts resulted in fluorescence of sporozoites within whole oocysts that were still capable of excystation, while oocysts that were dead when fixed only fluoresced at background levels. The decline in excystation seen with oocysts over time in storage correlates well with the loss of FISH signal (11). One significant advantage of using FISH for viability determinations is that samples can be fixed after concentration enabling viability to be subsequently determined. This ability to retrospectively determine viability has advantages in terms of sample processing and biosafety.

However, it needs to be emphasised that FISH is an excellent method to examine oocysts as long as a potential reduction or loss of viability is due to the 'natural cause' of ageing and prolonged exposure to the environment. Problems arise when oocysts are examined that have been inactivated through the application of disinfectants, heat or irradiation of oocysts. These inactivation methods will leave the relatively stable rRNAs intact and when subjected to FISH the oocyst, although inactivated, will appear viable. The limitation may be overcome by the addition of RNAse which degrades the target rRNA in non-viable oocysts (Gunasekera pers. comm.). The precondition is that the method applied for oocyst inactivation causes a certain degree of permeabilisation allowing RNase molecules to penetrate the oocyst wall.

FISH analysis is compatible with currently used IFA protocols. We have recently developed two-step staining procedure, which takes less than 40 minutes to complete, that encompasses staining with FITC-labelled (green) monoclonal antibodies, Texas Red labelled FISH probes and DAPI (blue) to give a 3-colour confirmation step.

4 MOLECULAR BEACONS

Oligonucleotide probes used for FISH can bind non-specifically to debris in water samples resulting in increased background fluorescence. This becomes problematical if the FISH signal from target organisms is weak. Molecular beacons are oligonucleotide probes originally designed to detect the presence of specific nucleic acid targets in homogenous solutions (12). They were shown to be useful in situations where it was not possible or desirable to separate probe-target hybrids from an excess of unbound hybridisation probes that would subsequently cause high background signal problems.

Molecular beacons function similar to conventional nucleic acid probes but are modified in order to eliminate signals from unbound or non-specifically bound probes. The core of a molecular beacon is a specific sequence directed against a target of choice. Overhang sequences are then added to both the 5'- and the 3'-end of the specific sequence. These overhang sequences usually comprise 5 nucleotides each and are complementary to each other. The 5'-end of the beacon is conjugated to a fluorochrome. Molecular beacons can be synthesized in a variety of colours, thus enabling the detection of multiple targets in the same solution by utilizing a broad range of different fluorochromes (13). The 3'-end is modified by conjugation to a quencher molecule. The first step of this modification is the addition of a 3'-end primary amino group to the oligonucleotide which is the coupled with DABCYL (4-(4'-diethyl-aminophenylazo)benzoic acid) succinimidyl ester). Under non-denaturing conditions, due to the complementary overhang sequences, beacons that are not attached to their specific targets will form a 'hairpin' structure (Figure 2 A). As a result, the fluorochrome and the DABCYL are brought in close proximity to each other. Any excitation light will not cause the fluorochrome to emit a signal as the energy is absorbed by the quencher molecule through a phenomenon called 'fluorescence resonance energy transfer' (FRET). Subsequently, unbound probe will not generate a background signal and there is no need for the removal of excess probe. However, when bound to a target sequence the fluorochrome and quencher are separated resulting in a fluorescent signal (Figure 2B).

It was recognized earlier that molecular beacons can be useful tools for applications other than standard or multiple hybridisations in homogeneous solutions where removal of excess probe is either not feasible or desirable. For *Cryptosporidium* detection, FISH can be problematic due to non-specific binding of fluorescently labelled probes to detritus in environmental samples. Molecular beacons offer the prospect of overcoming this limitation due to the fact that they only fluoresce when bound specifically to their target.

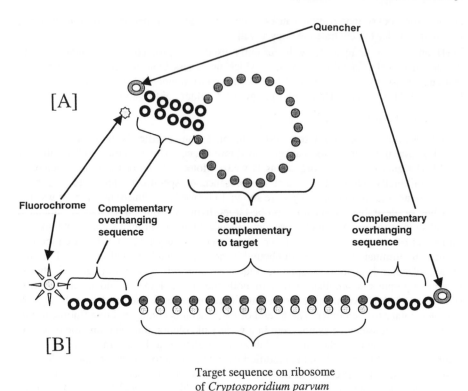

Figure 2 Molecular Beacons for *Cryptosporidium*. [A] When beacons are not attached to target under non-denaturing conditions the complementary overhanging sequence causes the formation of a hairpin structure bringing the quencher and fluorophore together. When in close proximity to the quencher fluorescence energy transfer from the fluorophore results in no fluorescence. [B] When attached to the target sequence within *Cryptosporidium* the hairpin structure is lost, separating quencher and fluorophore and the beacon will fluoresce

Design of molecular beacons: A basic rule for the design of molecular beacons relates to the melting points (T_m) of the hybrid formed by the specific probe sequence and its target, and the T_m of the complementary overhang sequences that enable formation of a hairpin structure under non-denaturing conditions. Two methods can be used for T_m calculation of probes. For probes with a length of 20 nucleotides the simple formula 2°C for each A and C residue and 4°C for each G and C residue can be applied. However, the method should not be applied for oligonucleotides that are either significantly shorter or longer than 20 residues as the calculated T_m will become increasingly inaccurate. Oligonucleotides longer than 20 residues are subjected to T_m calculation applying the 'nearest neighbour' formula (14).

The conventional versions of probes Cry-1 and Cry-2 both have a rather low T_m, which poses a problem for the design of these probes as molecular beacons. It is a precondition for the functionality of beacons that the hybrid of probe and target is the preferred and thermally more stable compared to the hairpin structure that can be formed by the complementary overhangs. If this were the case the hairpin structure would be the preferred conformation of the probe and no hybridisation to a specific target will occur.

The melting points of the complementary overhangs need to be determined and must not exceed the melting point of the probe/target hybrid. None of the above methods for calculation of the T_m of a given probe is suitable for the overhangs as these typically consist of only 5 nucleotides. The program RNAdraw (15) was used to calculate overhang melting points in order to determine overhang sequences that are compatible with the specific probe sequences and enable probe/target hybrid formation.

Preliminary results:

Both molecular beacons Cry 1 and Cry 2 were tested both individually and in combination using standard hybridisation protocols. When used individually both showed satisfactory results (Figure 3). The signal intensity measured on an arbitrary scale reaching from 0 – 250 was similar to that of conventional FISH probes although the background level of fluorescence was considerably reduced. An unexpected result was obtained when both beacons were used in combination. The signal intensity decreased by more than 30 % compared to hybridisations employing only one of the beacons (Figure 3.)

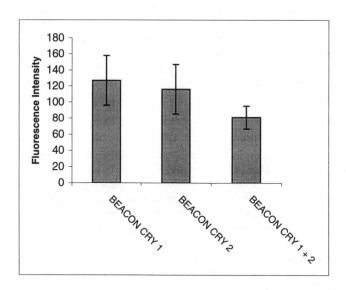

Figure 3   Application of molecular beacons Cry 1 and Cry 2 to the detection of *Cryptosporidium* oocysts.

When non-beacon FISH probes Cry 1 and Cry 2 are used in combination fluorescence signal increases. The reason for this significant decrease in signal strength with beacons is not clear. However, it seems likely that the quencher molecules of the probes interfere with each other's fluorochrome. The target sites of Cry-1 and Cry-2 on the 18S rRNA are separated by a relatively short stretch of approximately 100 nucleotides. The three dimensional structure of the 18S rRNA within the ribosome might cause even closer proximity of the two target sites.

Problems associated with molecular beacons:

There are a number of basic rules for the design of molecular beacons. These relate to the length of beacons and the complementary overhang sequences, the G+C content ratios and the differences in melting points of specific sequences and overhangs. However, strict observation of the basic rules does not necessarily guarantee a functional product. In terms of molecular beacons, functionality means efficient binding to the target and elimination of background signals through optimal quenching. We experimented with a variety of molecular beacons and although all were designed according to the same criteria the results were very different. In many cases efficient binding and background reduction were achieved but occasionally the degree of non-specific binding of a beacon to contaminating particles was so great that identification of target cells was not possible. In these cases the quenching appeared to be inefficient indicating that no secondary structure 'hairpin' formation occurred. Attempts to rectify the problem by changing the length of the overhang sequences and/or the T_m ratios between specific sequences and overhangs were unsuccessful.

These results demonstrated an element of uncertainty in the design of molecular beacons. Should the beacon versions of Cry-1 and Cry-2 require modification to improve background reduction we will consider employing synthetic target DNA. These artificial targets for molecular beacons are useful tools that allow spectrophotometrical monitoring of the degree of quenching, fluorometric response to the addition of target, the thermal denaturation profiles of a molecular beacon and the hybrid formed between a molecular beacon and its oligonucleotide target.

5 FLOW CYTOMETRY

Flow cytometry for the detection *Cryptosporidium* in water was first pioneered by Thames Water (16) and has been used for about 10 years to analyse water samples in the UK and Australia. Flow cytometry is used as a means of separating fluorescently labelled oocysts from other debris found in water concentrates using a technique called fluorescence activated cell sorting (FACS). Because FACS effectively purifies oocysts from other debris microscopy is much simpler, quicker and more reliable than using separation methods based on floatation (1)

Despite the high recovery efficiencies, speed of analysis and ease of confirming results, flow cytometry has not been extensively applied to routine monitoring for protozoan parasites outside the UK and Australia. This is due mainly to:

- high cost (>$US 200,000) of instruments such as the Coulter Elite, or Becton Dickinson (BD) FACSVantage.

- high maintenance costs (*ca.* $US 20,000 p.a.) of sophisticated cytometers that have cell sorting capabilities.
- loss of analysis time due to instrument downtime; particularly problems associated with instrument blockage.
- time and skill required to calibrate the instruments (*ca.* 1hour/day)
- requirement for a cytometrist.
- regulatory constraints on the methods that can be used in the USA.

BD FACSCalibur for *Cryptosporidium* analysis:

While the Coulter Elite and Becton Dickinson FACSVantage are complex research tools, simple to use automated flow cytometers, designed for routine clinical applications, are available. Such instruments:
- are reliable and robust.
- are simple to use, and do not require a specialist cytometrist.
- are factory preset, and thus do not require daily calibration.
- have low maintenance costs.
- are commonly found in clinical analysis laboratories.

One of these simple cytometers, the BD FACSCalibur, has a cell sorting option enabling oocysts to be selectively sorted for confirmation using techniques such as PCR, staining with fluorescent dyes (i.e. DAPI), or fluorescence *in situ* hybridisation (FISH).

The BD FACSCalibur (*ca.* $US 120,000) is less than half the price of a Coulter Elite or BD FACSVantage. However, the BD FACSCalibur is a lower performance instrument and using the commercially available reagents and standard instrument configuration no group has been able to obtain acceptable recoveries with turbid water samples without an additional clean up step such as density gradient centrifugation (17).

When analysing concentrated water samples for protozoan parasites using the BD FACSCalibur, we have identified 2 sources of losses (18). First cysts and oocysts were found to adhere to the sample tube. Second, the limited performance of the instrument was found to result in lower recoveries due to a problem known as event coincidence. The problem of oocysts sticking to the sample tube has been overcome by including a detergent washing procedure. Losses due to event coincidence have been overcome using the specific monoclonal antibodies CRY104 and G203 in conjunction with novel instrument modifications and settings specifically suited to environmental analysis. A full protocol for this method can be found at http://www.bio.mq.edu.au/flowgird/root/Slideshows/calibur/protocol.htm.

6 TWO-COLOUR 'ANALYSIS ONLY' FLOW CYTOMETRY

To date, all flow cytometric procedures for the detection of *Cryptosporidium* have used FACS to select oocysts from debris for subsequent microscopic confirmation. Most clinical applications of flow cytometry do not involve FACS. Rather simple to operate and relatively inexpensive 'analysis-only' instruments (eg BD FACSCount, Coulter XL), which lack cell-sorting capabilities are used for complete analysis without microscopic confirmation. Analysis-only instruments rely on the ability to optically distinguish target cells from background with a high degree of certainty. To achieve the level of specificity required more than one colour of fluorescence is necessary. The prospect exists of developing entirely analysis-only instrumentation for *Cryptosporidium* detection that would dispense with FACS and microscopic confirmation. Such instrumentation could potentially be portable or adapted for on-line *Cryptosporidium* detection.

Analysis-only detection requires that optical properties of oocysts be altered sufficiently to enable their discrimination, by flow cytometry, from other particles in a sample. Currently, oocysts are tagged using fluorescently labelled mAbs with one colour (typically green). Environmental samples contain autofluorescent particles such as minerals and algae that optically resemble fluorescently green stained oocysts. In addition, non-specific binding of fluorescent mAbs to detrital particulates results in increased fluorescent background. Hence, using one colour flow cytometry there is a need to sort suspect particles and confirm by microscopy.

Using another protozoan, *Dictyostelium discoideum,* Vesey and co-workers (19) have demonstrated the specific detection of a single spore in 5 litres of river water using two-colour analysis-only flow cytometry. Two highly specific non-competing surface mAbs, one labelled with fluorescein isothiocyanate (FITC; green-fluorescence) and the other labelled with phycoerythrin (RPE; orange-fluorescence) were used in this analysis. The use of two colours on different, highly specific non-competing mAbs provided the optical discrimination required for single spore detection in river water.

However, for *Cryptosporidium* all mAbs that are specific to the oocyst outer wall bind to the same immuno-dominant epitope and thus compete for the same binding sites (20). In the absence of non-competing mAbs various competing mAbs labelled with RPE and FITC have been evaluated (21). With competing mAbs two-colour staining did not provide the discrimination required for single oocyst detection. However up to a 100-fold reduction in the degree of non-target material sorted was demonstrated compared to one-colour analysis. The reduction in sorted detritus significantly improved analysis times by microscopy. Additionally, two-colour immunofluorescence aided the microscopic confirmation of suspect oocysts through the second (colour) identification parameter.

Two-colour antibody flow cytometric analysis post IMS:

To improve sensitivity further, immunomagnetic separation (IMS) has been utilised as a pre-enrichment step to reduce levels of contaminating detritus prior to two-colour flow cytometric analysis (22). Immunomagnetic separation combined with one-colour detection of oocysts resulted in an average of 20.5 non-*Cryptosporidium* and non-*Giardia* particles detected within 10 litre raw water samples. This was reduced to 5.6 non-*Cryptosporidium* and non-*Giardia* detects using two-colour analysis (Table 1).

Table 1 Number of non-target particles sorted from a 10 litre concentrated surface water samples using 1 and 2- colour flow cytometric analysis using either FITC (1-colour) or PE and FITC labelled Cry 104 (22).

Analysis method	Sort Total
1 Colour	20.5 ± 31.00
2 Colour	5.6 ± 3.00

Two-colour flow cytometric analysis involving FISH and antibody staining:

The lack of specific mAbs that recognise independent epitopes on the surface of *Cryptosporidium* is the major factor limiting the development of analysis-only flow cytometric detection methods. FISH could provide an alternative method for two-colour labelling of oocysts. FISH has the advantage over antibodies that probes specific to *C. parvum* are available. FISH provides information on viability and the target site (rRNA) is independent of the target site for antibodies. However using oligonucleotide probes we have found that the fluorescent signals is not sufficiently bright enough for FCM (9, 10). Recently, peptide nucleic acid (PNA) probes specific to *Giardia lamblia* and *Cryptosporidium parvum* have been developed and tested in our laboratory. When used in FISH procedures, PNA probes give significantly brighter fluorescent signals than equivalent oligonucleotide probes. The fluorescence intensity of PNA probes is similar to that obtained with fluorescent mAb staining of *Cryptosporidium* (Dorsch, pers. Comm.). Preliminary investigations combining two *Giardia*-specific PNA probes labeled with FITC and the IgG_1 mAb G203 directed to the cyst wall labelled with RPE were used to detect cysts. Fluorescence intensities of cysts stained with the PNA probe and mAb combination were significantly brighter than dual mAb stained cysts (Ferrari, pers. comm.). This increased fluorescence reduced the signal to noise compared with dual mAb staining resulting in sensitivities sufficient for the detection of a single *Giardia* cyst in backwash water samples equivalent to 10 litres.

7 CONCLUSION

In the future, by combining mAbs and FISH probes in a multi-colour FCM assay, the potential exists for online detection that provides additional information on viability and species of detected oocysts. The major limitation existing is the lack of a FISH hybridisation technique that can be carried out in water samples prior to FCM. Development of such an assay, followed by method validation in a range of water types is necessary before this analysis-only method can be routinely utilised.

8 REFERENCES

1.   G. Vesey, P. Hutton, A. Champion, N. Ashbolt, K. L Williams, A.Warton, D. A Veal,*Cytometry*. 1994, **16**, 1-6.
2.   J. L Clancy, Z.Bukhari, R.M. McCuin, Z.Matherson, C.R. Fricker, *American Water Works Association*. 1999, **91**, 60-68.
3.   P McClellan,. *Sydney Water Inquiry*. Sydney: NSW Premier's Department,1998.
4.   J. L Clancy,. *Journal of American Water Works Association*. 2000, **92**, 55-66.
5.   R. Hoffman, C. Chauret, J.Standridge, L. Peterson, *Journal of the American Water Works Association*. 1999, **91**, 69-78.
6.   G. Vesey, D. Deere, D. C. Weir N. Ashbolt, K. Williams, K. Griffith, D. Veal, *Letters in Applied Microbiology*. 1997, **25**, 316-320.
7.   B. C. Ferrari,. G. Vesey, C. Weir, K.L. Williams, D.A. Veal, *Water Research*. 1999, **33**, 1611-1617.
8.   Weir, C., Vesey, G., Slade, M., Ferrari, B., Veal, D.A., Williams, K. *Clinical and Diagnostic Laboratory Immunology*. 2000, **7**, 745-750.

9.   G. Vesey, N. Ashbolt, E. J.Fricker, D. Deere, K. L. Williams, D. A. Veal, Dorsch, M., *Journal of Applied Microbiology*. 1998, **85**, 429-440.
10.   D. Deere, G. Vesey, M. Milner, K. Williams, N. Ashbolt, D. Veal. *Journal of Applied Microbiology*. 1998, **85**, 807-818.
11.   G. Vesey, N. Ashbolt,. Wallner, M. Dorsch, K. Williams, D. A. Veal, 1995. In *Assessing Cryptosporidium parvum oocyst viability with fluorescent in-situ hybridisation using ribosomal RNA probes and flow cytometry*, ed. W. B. Betts, D. Casemore, C. Fricker, H. Smith, H. Watkins, pp. 131-138. Cambridge, UK: Royal Society for Chemistry
12.   S. Tyagi, F.R Kramer,. *Nature Biotechnology*. 1996, **14**, 303-308.
13.   S. Tyagi, D.P.Bratu, F.R Kramer,. *Nature Biotechnology*. 1998, **16**, 49-53.
14.   K. J.Breslauer, F. Shaffer, H. Blöcker, L.A. Marky, *Proceedings of the National Academy of Sciences of the USA*. 1986, **83**, 3746 - 3750.
15.   O. Matzura, A. Wennborg, *Computer Applications in the Bioscience*. 1996, **12**, 247 - 249.
16.   G. Vesey, J. S.Slade, C. R. Fricker, *Letters in Applied Microbiology*. 1991, **13**, 62-65.
17.   G. J. Medema, F. M. Schets, H. A. M. Ketelaars, G. Boschman, *Water Science and Technology*. 1998, **38**, 61 - 65.
18.   M. Gauci, S. Le Moenic, G. Vesey, D. Deere, K. L. Williams, J. Piper, D. A. Veal, 1997. *Methods of detection of Cryptosporidium parvum oocysts in environmental water samples using the Becton Dickinson FACSCalibur sorting flow cytometer.* Presented at the 10th International Congress of Protozoology, Sydney, 21 - 25 July 1997.
19.   G. Vesey, J. Narai, N. Ashbolt, K.Williams, D. A. Veal, 1994. *Detection of Specific Microorganisms in Environmental Samples using Flow Cytometry*, ed. Z. Darzynkiewicz, P. Robinson, pp. 488-521. New York: Academic Press.
20.   A. G. Moore, G. Vesey, A. Champion, P. Scandizzo, D. Deere, D.A. Veal, K.L Williams, *International Journal of Parasitology*. 1998, **28**, 1205-1212.
21.   B. C. Ferrari, G. Vesey, K.A. Davis, M. Gauci, D. Veal, A novel two -colour flow cytometric assay for the detection of *Cryptosporidium* in environmental samples. *Cytometry*. 2000, **41**, 216-222.
22.   Ferrari, B. C. In *Dual-colour flow cytometric detection of Cryptosporidium and Giardia in water.* Sydney: Macquarie University, 2000. p277.

CAN WE BELIEVE OUR RESULTS?

Frank W. Schaefer, III

National Environmental Research Laboratory
U.S. Environmental Protection Agency
26 West M.L. King Drive
Cincinnati, Ohio 45268, USA

1 INTRODUCTION

Numerous waterborne outbreaks of cryptosporidiosis have occurred recently with the most notable being the 1993 episode in Milwaukee. Due to these outbreaks and the concern for public health, the past decade has seen a massive effort expended on the development of methods to detect *Cryptosporidium parvum* oocysts in source and finished water. Initial analyses for *C. parvum* oocysts were based on methods developed for detecting *Giardia* cysts in water. In short, the procedure involved sampling either 100 liters of source water or 1,000 liters of finished water using a fiber wound filter with a nominal porosity of 1.0 µm. After washing the filter fibers with buffer, the extracted particulates containing the parasites were further concentrated by centrifugation. Buoyant density centrifugation was then used to separate the parasites from the particulates. Because buoyant density centrifugation is not a selective purification procedure, the parasites still were associated with and masked by particulates. To overcome this problem, the parasites were selectively stained with fluorescently labeled monoclonal antibodies. The results of the assay were microscopically determined using epifluorescence microscopy to detect the presumptive parasites and differential interference contrast microscopy to confirm the identity of the parasites by demonstration of internal morphological characteristics. In the case of *Cryptosporidium*, recovery rates ranged from 0 to 11% in seeded samples.

Method 1622 for *Cryptosporidium* was developed using off the shelf reagents and equipment in an attempt to improve on oocyst recovery rates. Instead of sampling large volumes of water, 10 liter samples are concentrated by passing the water through a 1.0 µm absolute porosity filter. Immunomagnetic separation, a more selective procedure employing monoclonal antibodies, is used in place of buoyant density centrifugation to separate the parasites from the particulates. Like the previous method, staining is done with fluorescently labeled monoclonal antibodies for detection. However, in addition to differential interference contrast microscopy for confirmation, the samples are counter stained with 4',6-diamidino-2-phenylindole (DAPI) which helps to demonstrate sporozoite nuclei. Even with these improvements, the recovery results for the Method 1622 only average around 38%.

2 LABORATORY PERSONNEL AND EQUIPMENT

The competence of the laboratory personnel and the quality of the laboratory equipment has a pronounced influence on oocyst recovery. In an attempt to overcome and control these variables, performance evaluation samples and laboratory site visits have been instituted in Australia and the United Kingdom. Similar measures also were used in United States. Out of 72 laboratories that applied to analyze Information Collection Rule samples in the United States, only 35 laboratories were approved. Results from a study of laboratory proficiency conducted by the U.S. Environmental Protection Agency in anticipation of the Information Collection Rule are shown in Figure 1. These data show that there is significant variation both within and between laboratories. By comparing Figure 1 to Figure 2, one can see that laboratories that performed well in detecting *Cryptosporidium* oocysts also performed well in detecting *Giardia* cysts. Surprisingly in this study some of the most well known laboratories produced the poorest results.

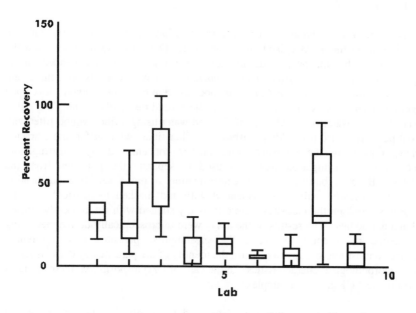

Figure 1 *Percent Recovery of* Cryptosporidium *by Laboratory*

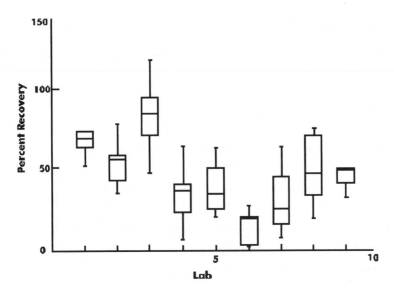

Figure 2 *Percent Recovery of* Giardia *by Laboratory*

To date in Australia, only three out of eight laboratories are accredited by the National Association of Testing Authorities, Australia, to analyze water samples for *Giardia* cysts and *Cryptosporidium* oocysts. Laboratories in the United Kingdom are doing better, as there are 14 accredited labs for the regulatory *Cryptosporidium* analysis. Although none of the applicants were ever rejected, some of the laboratories took three or four inspections to reach an acceptable level of performance. This suggests that consumers of laboratory services need to be cautious. The consumer should always ask to see internal quality assurance data, like initial and ongoing performance and recovery data, as well as whether the laboratory routinely analyzes blind proficiency samples.

A major reason that laboratories have not been accredited was because the analysts were not able to correctly analyze performance evaluation samples and reported either false positives or negatives or reported percent recoveries out of the acceptance range. Other reasons included improper quality assurance, improper performance of the analytical protocol, and not utilizing the prescribed equipment and/or reagents.

3 NUMBER OF PROCESSING STEPS

As illustrated in Table 1, the greater the number of processing steps in a method the lower the oocyst recovery is. One can see that by the time all the processing steps have been completed only 31 out of the 100 oocysts remained. Although Table 1 is just a simulation, the recovery value is similar to the average 35 % oocyst recovery reported for spiked reagent water and 43% for spiked raw surface water reported for Method 1622 as part of a 12 laboratory round-robin validation study.[1] The Method 1622 validation study was carried out using some of the most qualified laboratories in the United States. Lesser qualified laboratories would be expected to have percent recoveries much lower or to even report false negatives. Laboratory skill usually improves, after having processed numerous samples, and this usually results in an improvement in the recovery results. Water quality parameters like turbidity and algal content can skew the recovery data up or down depending on quantities present.

Table 1 *Effect of Processing on Oocyst Recovery*

Processing Stages	Processing Step Oocyst Recovery (%)	Oocyst Number
Initial oocyst spike	100	100
Sampling	85	85
IMS	60	57
Staining	66	37
Reading	85	31

4 COMMERCIAL PRODUCTS AND REAGENTS

Most methods are dependent on commercially prepared products and reagents like *C. parvum* oocysts, antibody stains, immunomagnetic separation kits and filters. Current practice calls for *C. parvum* oocysts to be used in quality assurance and control studies associated with all recognized *Cryptosporidium* oocyst detection methods. Furthermore, demand has increased significantly over the past ten years for oocysts needed to study physical removal and chemical inactivation of these parasites. It has become increasing apparent from reviews of the first studies using oocysts that little if any care was exercised as to the source or isolate of *Cryptosporidium* used, the way in which the oocysts were prepared, how they were stored, and the age of the oocysts from the time of passage and purification to when they were used. Perhaps even more distressing is the fact that early researchers had no acceptance criteria for the quality of oocysts used in their studies. In many cases, the early control studies were conducted by individuals who had little or no training in microbiology. As a result they were totally reliant on the labeling supplied by the oocyst vendors. Because of their lack of knowledge in this area, they did not have the ability to use a microscope for estimating oocyst quality and densities with a hemocytometer. Often the numbers of oocysts listed on the container were not as high as reported by the vendor due to sticking and deterioration during shipment. Consequently, the individuals had no idea what they actually used in their studies. There even are anecdotal reports of the wrong organism being shipped by mistake to investigators.

Classical approaches to obtaining oocysts are based not so much on purification as on concentration for clinical identification. Basically, two concentration approaches routinely have been used: buoyant density flotation and formalin ether sedimentation. In the case of the latter procedure, either ether or ethyl acetate is used to remove lipids from the fecal sample. In such a procedure lipids will also be removed from any parasites present and change the parasite's surface characteristics.[2] Because most coccidian parasites must be sporulated, dogma calls for placing fecal material immediately into potassium dichromate to retard microbial growth. It is now known that the zeta potential on the surface of oocysts is significantly changed by exposure to potassium dichromate and extremes of pH.[3,4] Some techniques for purification as well as for excystation of coccidian oocysts call for exposure to household hypochlorite solutions. Reduker et al.[2] published a transmission electron microscopy study in which they clearly showed the outer layer of the *C. parvum* oocyst being stripped off by exposure to hypochlorite. From the studies above, it is now known that the way in which oocysts are treated and purified from fecal material has a pronounced impact on their utility for water treatment research. For example, any oocysts purification step that changes either the viability and/or oocyst wall will impact the way these organisms respond to antibodies, to flocculating agents, or to disinfectants.

Immunomagnetic separation and immunofluorescent staining kits are antibody based and have been shown to have significant lot-to-lot variation by Hoffman et al.[5] They used the flow cytometer to measure the lot-to-lot variability, cross reactivity, and avidity of four commercially available staining kits. All of the antibody kits showed acceptable avidity reactions with *Giardia* cysts, while two of the kits formed low avidity complexes with *Cryptosporidium* oocysts. Lot-to-lot variability was shown to be greater in the case of two of the four kits. Specificity problems were detected in that two of the staining kits reacted with non-*parvum Cryptosporidium*, and two of the kits cross-reacted with *G. muris* in addition to *G. lamblia* cysts. *Microcystis aeruginosa* and *Aphanocapsa rivularis*, which are cyanobacteria in the same size range as *C. parvum* oocysts, cross-reacted with one of the anti-*Cryptosporidium* antibodies. Consequently, identification of *C. parvum* oocysts in natural water samples solely on the basis of immunofluorescent staining of an organism of the right size and shape is unacceptable due to the antibodies cross reacting with other waterborne microorganisms of the same size and shape. Sample processing steps, like calcium carbonate flocculation and formalin fixation, also were shown to diminish antibody staining. All this information points to and reemphasizes the need for a scrupulous quality assurance program and the need for selecting the best antibody for the task at hand. Laboratories routinely should test each lot of antibody before using it for actual experimental work.

Confounding detection and identification of *Cryptosporidium* oocysts in water samples are all the autofluorescent inanimate objects and fluorescent chloroplast-containing algae. Masking of oocysts also occurs, when immunomagnetic or buoyant density purification procedures have not been optimal. Therefore, positive identification of *C. parvum* oocysts must include either demonstration of internal morphological characteristics or a positive genetic probe reaction. It should be noted, however, that both the internal morphological characteristics and molecular probe reactions can be misinterpreted.

The filter porosity in protozoan analysis of water samples is usually selected on the basis of the smallest organism. *Cryptosporidium parvum* oocysts are 4-6 µm in size and are known to be plastic or flexible enough to squeeze through 2 μm pores.[6] With 1μm nominal porosity filters, which have a porosity range from sub-micron up to 30 μm, the majority of the oocysts pass right through the filter. In the case of 3 μm absolute porosity filters, which have been suggested for some analytical protocols, some of the oocysts will be able to pass through the filter. Because oocysts can pass through a 2 μm but not 1 μm absolute porosity filter, 1 μm absolute porosity filters are the filters of choice. Scanning electron microscopy studies of polycarbonate track-etched filters have shown that the pores are not equidistant. In fact, some of the pores were shown to not only be adjacent but to have been so closely spaced that they created a pore twice the specified size.   Capsule filters are known to have lot-to-lot failures.[7] Furthermore, compressed foam filters appear to have a shelf life. If not used within 9 to 12 months of manufacture, they sometimes fails to expand and wash properly during the elution of the filter.   All these deviations in product quality point to the need on part of the manufacturers to institute better quality control.

5 PHYSICAL AND MICROBIOLOGICAL FACTORS IN WATER

In environmental water samples there are physical and microbiological factors which can compromise the epitopes against which the antibodies react, and as a result *C. parvum* oocysts will not be detected. Moreover, the matrix may possess as yet undefined parameters which, while not compromising the epitope on the parasite, will compromise formation of the antibody-epitope complex. Indeed, if this happens, then both the immunomagnetic separation and the immunofluorescent staining will be unsatisfactory. Until the manner in which various matrix effects interact with the immunochemical reactions associated with this methodology are critically defined, progress in overcoming them cannot be made.

6 OOCYST DISTRIBUTION

Oocysts are not always uniformly distributed throughout a water sample. If the whole sample is not analyzed, then fallacious results can occur. Similarly, oocysts are not present in source and finished water at all times. Experience indicates that the probability of finding oocysts increases after rain events. This creates a dilemma, for the increased matrix components may reduce recovery. However, when infrequent samples are taken, the probability of detection is diminished.

7 SUMMARY

Keeping all of this information in mind, can we believe our results - probably not. Falling prey to any of these sources of error yields quantitative results little better than randomly generated numbers.   The only meaningful result is a positive detection, which is likely to be a gross underestimate of the true value.

References

1. U.S. Environmental Protection Agency. Interlaboratory validation study results for *Cryptosporidium* precision and recovery for U.S. EPA method 1622, Washington D.C., 1999.

2. Reduker D.W., C.A. Speer, and J.A. Blixt. Ultrastructural changes in the oocyst wall during excystation of *Cryptosporidium parvum* (Apicomplexa: Eucoccidorida). *Canad. J. Zool.*, 1985, **63**, 1892-6.

3. Brush, C.F., M.F. Walter, L.J. Anguish, and W.C. Ghiorse. Influence of pretreatment and experimental conditions on electrophoretic mobility and hydrophobicity of *Cryptosporidium parvum* oocysts. *Appl. Environ. Microbiol.*, 1998, **64**, 4439.

4. Drodz, C. and J Schwartz. Hydrophobic and electrostatic cell surface properties of *Cryptosporidium parvum*. *Appl. Environ. Microbiol.*, 1996, **62**, 1227-32.

5. Hoffman, R.C., C. Chauret, J. Standridge, and L. Peterson. Evaluation of four commercial antibodies. *J. Am. Wat. Wks. Assoc.*, 1999, **91**, 69-78.

6. Li, S. *Cryptosporidium* potential surrogates and compressibility investigation for evaluating filtration based water treatment technology. Masters Thesis, Miami University, Oxford, Ohio, 1994.

7. Simmons, III, O.D., M.D. Sobsey, C.D. Heaney, F.W. Schaefer III, and D.S. Francy. Concentration and detection of *Cryptosporidium* oocysts in surface water samples by Method 1622 using ultrafiltration and capsule filtration. *Appl. Environ. Microbiol.*, 2001, **67**, 1123-7.

Subject Index